Progress in Community Nutrition: Dietary Patterns and Planetary Health

Progress in Community Nutrition: Dietary Patterns and Planetary Health

Editors

Colin Bell
Penelope Love

MDPI • Basel • Beijing • Wuhan • Barcelona • Belgrade • Manchester • Tokyo • Cluj • Tianjin

Editors
Colin Bell
Deakin University
Australia

Penelope Love
Deakin University
Australia

Editorial Office
MDPI
St. Alban-Anlage 66
4052 Basel, Switzerland

This is a reprint of articles from the Special Issue published online in the open access journal *Nutrients* (ISSN 2072-6643) (available at: https://www.mdpi.com/journal/nutrients/special_issues/Community_Nutrition_Dietary_Patterns).

For citation purposes, cite each article independently as indicated on the article page online and as indicated below:

LastName, A.A.; LastName, B.B.; LastName, C.C. Article Title. *Journal Name* **Year**, *Volume Number*, Page Range.

ISBN 978-3-0365-6848-5 (Hbk)
ISBN 978-3-0365-6849-2 (PDF)

Cover image courtesy of Colin Bell

Contents

About the Editors

Colin Bell

Colin Bell leads the Public Health Medicine theme in the Doctor of Medicine course at Deakin University. Colin is recognized for his expertise in obesity, nutrition and chronic disease prevention and has worked on a number of successful population and community-based programs in Australia, New Zealand and the Pacific.

Penelope Love

Penny is an Advanced Accredited Practicising Dietitian (Adv APD), and senior lecturer for Public Health and Community Nutrition within the School of Exercise and Nutrition Sciences (SENS) at Deakin University. Penny is a member of the Institute for Physical Activity and Nutrition (IPAN), researching the translation, implementation and sustainability of early childhood prevention at scale to inform the identification of key leverage points within existing service delivery systems.

Preface to "Progress in Community Nutrition: Dietary Patterns and Planetary Health"

This Special Issue contains original research that describes comprehensive approaches to nutrition surveillance, epidemiological studies of diet, the development, implementation, and evaluation of dietary recommendations and goals, interventions to improve access and availability of healthy foods and interventions to strengthen food literacy, and sustainable dietary patterns among populations and health professionals. It targets researchers and policymakers working to make healthy and environmentally food choices easier. We thank the participants and researchers that made the special issue possible.

Colin Bell and Penelope Love
Editors

Editorial

Community at the Centre of Future Food Systems

Colin Bell [1],* and Penelope Love [2],*

[1] Institute for Health Transformation, School of Medicine, Deakin University, Geelong, VIC 3216, Australia
[2] Institute for Physical Activity and Nutrition, School of Exercise and Nutrition Sciences, Deakin University, Geelong, VIC 3216, Australia
* Correspondence: colin.bell@deakin.edu.au (C.B.); penny.love@deakin.edu.au (P.L.)

Citation: Bell, C.; Love, P. Community at the Centre of Future Food Systems. *Nutrients* **2022**, *14*, 4951. https://doi.org/10.3390/nu14234951

Received: 4 November 2022
Accepted: 9 November 2022
Published: 22 November 2022

Publisher's Note: MDPI stays neutral with regard to jurisdictional claims in published maps and institutional affiliations.

Have you thought about what it is you love about food? It is likely that it is not as much about the nutrients in the food or even how tasty the food is as it is about the circumstances in which your food was sourced, prepared, and eaten, and the connections you have with the people involved.

Indigenous knowledge systems teach us that connections with nature and culture are central to our identity and that food is central to those connections [1]. Food connects us to the land, oceans and sky that sustain us, as well as to our cultures. Food helps us know where we are from, 'who we are' and to 'feel good about ourselves' [2].

Community is the level of society where food-related connections to people and place occur. As such, community needs to be at the centre of future food systems in the same way food markets are at the centre of our villages, towns and cities. Increased interaction between farmers and consumers may help to achieve more environmentally sustainable food supply chains [3]. It may also reduce the stranglehold that global food manufacturers have on food supply and decrease the supply of ultra-processed foods [4]. Community, school and home gardens will improve food literacy and accessibility, and the affordability of healthy foods [5]. Where people are separated from community and culture due to climate change, conflict or economic opportunities, food can help bring them back together [6].

Globally, countries are pushing back on corporate and climate change disruptions to food systems and are looking for ways to strengthen community involvement in food systems, so that healthy and environmentally friendly food choices are easier to make. To achieve this, policymakers need access to the latest evidence on evolving dietary patterns and on effective solutions, and this Special Issue on Progress in Community Nutrition: Dietary Patterns and Planetary Health provides some of this evidence.

Using 2015 data from over 20,000 Canadians, Shafiee et al. identified cooked regular lean or extra-lean ground beef and white rice as the most frequently consumed animal-based and plant-based foods, respectively [7]. Based on a literature review, Gibbs and Cuppuccio identify meat appreciation, health concerns, convenience, and expense as barriers to transitioning to more plant-based food systems, which have a smaller environmental footprint than animal-based food systems [8]. Using unique data from the School Fruit and Vegetable Scheme in European Union Member States, Comino et al. found that over a 7-year period (2009/10 to 2016/17), it was possible for school children to receive up to 20 kg of fruit or vegetables delivered over 150 days [9]. However, not every child targeted by the scheme ended up with fruit or vegetables in their hands, and delivery days and the amount of fruit and vegetables provided varied considerably across countries. These studies illustrate the challenges faced by communities to transition to more environmentally friendly dietary patterns.

With the development of future dietary guidelines for Australia in mind, Wingrove et al. [10] point to the need for the inclusion of evidence on environmental sustainability and equity, in addition to evidence on dietary exposure and health. Research from Japan, led by Kurisaki and Kushida [11], and from Vietnam, led by Thien Mai [12], highlight how difficult dietary exposures are to measure, and of the need for quality and validated measurement techniques. The contribution from Du Plessis and colleagues reminds us that the translation of dietary evidence and guidelines into messages suitable for different target audiences is as important as the evidence itself [13]. Lastly, Swart et al., with their unique study on the dietary intake and diversity of landfill waste-pickers in South Africa, urges us not to neglect the nutritional status of vulnerable communities [14].

If community is more centrally engaged in the food system, it is possible that naturally occurring and culturally appropriate foods can be distributed more equitably, with an associated reduction in health disparities. Zoe et al. note in their scoping review that Chinese immigrants to Canada and the United States consumed less healthful diets (for example, higher caloric intakes and higher meat and alternative intakes) with greater acculturation [15]. Importantly, limited access to traditional or healthy foods was associated with acculturation. They note that 'decisions regarding policies on immigration and neighbourhood planning can also consider the health benefits associated with areas of high immigrant density, improve access to ethnic foods, and promote sharing of cultural food practices' [15]. Akbar and colleagues used a taro plant to illustrate themes that emerged from interviews with diverse representatives of Māori and Pasifika diaspora communities in Queensland, Australia [16]: 'Identity, hospitality and reciprocity, and spirituality form the roots or foundation of the meaning of food for Māori and Pasifika peoples; solutions and barriers form the soil or structure from which these understandings and practices of food are nourished or restricted and; physical and mental health, expectations and obligations, and stigma and shame form the leaves of the taro plant, representing the surface-level experiences and challenges of food insecurity' [16]. The authors conclude that 'when foods (including cultural foods) are not available or accessible, individuals experience not only a loss of physical health but lose connections to their broader identity and their link to their ancestral homes' [16].

The EAT–Lancet Commission framework for identifying 'win-win' diets (healthy and environmentally sustainable) was proposed as 'universal for all food cultures and production systems in the world, with a high potential of local adaptation and scalability' [17]. This Special Issue reveals that such local adaption is best achieved by positioning the community at the centre of future food systems.

Author Contributions: Conceptualization: C.B.; reference selection: C.B. and P.L.; writing—original manuscript: C.B. and P.L.; writing—review and editing: C.B. and P.L. All authors have read and agreed to the published version of the manuscript.

Conflicts of Interest: The authors declare no conflict of interest.

References

1. Pascoe, B. *Dark Emu*; Griffin Press: Melbourne, Australia, 2018.
2. Kuhnlein, H.V.; Erasmus, B.; Spigelski, D.; Burlingame, B. *Indigenous Peoples' Food Systems And Wellbeing: Interventions and Policies for Healthy Communities*; FAO: Rome, Italy, 2013.
3. Loiseau, E.; Colin, M.; Alaphilippe, A.; Coste, G.; Roux, P. To what extent are short food supply chains (SFSCs) environmentally friendly? Application to French apple distribution using Life Cycle Assessment. *J. Clean. Prod.* **2020**, *276*, 124166. [CrossRef]
4. Wood, B.; Williams, O.; Nagarajan, V.; Sacks, G. Market strategies used by processed food manufacturers to increase and consolidate their power: A systematic review and document analysis. *Glob. Health* **2021**, *17*, 17. [CrossRef] [PubMed]
5. Brennan, S.F.; Lavelle, F.; Moore, S.E.; Dean, M.; McKinley, M.C.; McCole, P.; Hunter, R.F.; Dunne, L.; O'Connell, N.E.; Woodside, J.V.; et al. Food environment intervention improves food knowledge, wellbeing and dietary habits in primary school children: Project Daire, a randomised-controlled, factorial design cluster trial. *Int. J. Behav. Nutr. Phys. Act.* **2021**, *18*, 23. [CrossRef] [PubMed]
6. Reddy, G.; van Dam, R.M. Food, culture, and identity in multicultural societies: Insights from Singapore. *Appetite* **2020**, *149*, 104633. [CrossRef] [PubMed]
7. Shafiee, M.; Islam, N.; Ramdath, D.D.; Vatanparast, H. Most Frequently Consumed Red/Processed Meat Dishes and Plant-Based Foods and Their Contribution to the Intake of Energy, Protein, and Nutrients-to-Limit among Canadians. *Nutrients* **2022**, *14*, 1257. [CrossRef] [PubMed]
8. Gibbs, J.; Cappuccio, F.P. Plant-Based Dietary Patterns for Human and Planetary Health. *Nutrients* **2022**, *14*, 1614. [CrossRef] [PubMed]

9. Comino, I.; Soares, P.; Martínez-Milán, M.A.; Caballero, P.; Davó-Blanes, M.C. School Fruit and Vegetables Scheme: Characteristics of Its Implementation in the European Union from 2009/10 to 2016/17. *Nutrients* **2022**, *14*, 3069. [CrossRef] [PubMed]
10. Wingrove, K.; Lawrence, M.A.; Russell, C.; McNaughton, S.A. Evidence Use in the Development of the Australian Dietary Guidelines: A Qualitative Study. *Nutrients* **2021**, *13*, 3748. [CrossRef] [PubMed]
11. Kurisaki, R.; Kushida, O. Number of Days Required to Estimate Habitual Vegetable Variety: A Cross-Sectional Analysis Using Dietary Records for 7 Consecutive Days. *Nutrients* **2022**, *14*, 56. [CrossRef] [PubMed]
12. Mai, T.M.T.; Tran, Q.C.; Nambiar, S.; Pols, J.C.V.d.; Gallegos, D. Development and Validation of the Vietnamese Children's Short Dietary Questionnaire to Evaluate Food Groups Intakes and Dietary Practices among 9–11-Year-Olds Children in Urban Vietnam. *Nutrients* **2022**, *14*, 3996. [CrossRef] [PubMed]
13. Du Plessis, L.M.; Job, N.; Coetzee, A.; Fischer, S.; Chikoko, M.P.; Adam, M.; Love, P.; on behalf of the Food-Based Dietary Guideline Technical Working Group Led by Tanzania Food and Nutrition Centre. Development and Field-Testing of Proposed Food-Based Dietary Guideline Messages and Images amongst Consumers in Tanzania. *Nutrients* **2022**, *14*, 2705. [CrossRef] [PubMed]
14. Swart, E.C.; van der Merwe, M.; Williams, J.; Blaauw, F.; Viljoen, J.M.M.; Schenck, C.J. Nutritional Status, Dietary Intake and Dietary Diversity of Landfill Waste Pickers. *Nutrients* **2022**, *14*, 1172. [CrossRef] [PubMed]
15. Zou, P.; Ba, D.; Luo, Y.; Yang, Y.; Zhang, C.; Zhang, H.; Wang, Y. Dietary Characteristics and Influencing Factors on Chinese Immigrants in Canada and the United States: A Scoping Review. *Nutrients* **2022**, *14*, 2166. [CrossRef] [PubMed]
16. Akbar, H.; Radclyffe, C.J.T.; Santos, D.; Mopio-Jane, M.; Gallegos, D. "Food Is Our Love Language": Using Talanoa to Conceptualize Food Security for the Māori and Pasifika Diaspora in South-East Queensland, Australia. *Nutrients* **2022**, *14*, 2020. [CrossRef] [PubMed]
17. Willett, W.; Rockström, J.; Loken, B.; Springmann, M.; Lang, T.; Vermeulen, S.; Garnett, T.; Tilman, D.; DeClerck, F.; Wood, A.; et al. Food in the Anthropocene: The EAT Lancet Commission on healthy diets from sustainable food systems. *Lancet* **2019**, *393*, 447–492. [CrossRef] [PubMed]

Article

Most Frequently Consumed Red/Processed Meat Dishes and Plant-Based Foods and Their Contribution to the Intake of Energy, Protein, and Nutrients-to-Limit among Canadians

Mojtaba Shafiee [1], Naorin Islam [1], D. Dan Ramdath [2,*] and Hassan Vatanparast [1,3,*]

[1] College of Pharmacy and Nutrition, University of Saskatchewan, Saskatoon, SK S7N 4Z2, Canada; mojtaba.shafiee@usask.ca (M.S.); naorin.islam@usask.ca (N.I.)
[2] Guelph Research and Development Centre, Agriculture and Agri-Food Canada, Guelph, ON N1G 5C9, Canada
[3] School of Public Health, University of Saskatchewan, Saskatoon, SK S7N 4Z2, Canada
* Correspondence: dan.ramdath@agr.gc.ca (D.D.R.); vatan.h@usask.ca (H.V.);
Tel.: +1-226-217-8082 (D.D.R.); +1-306-966-8866 (H.V.); Fax: +226-217-8181 (D.D.R.); +306-966-6377 (H.V.)

Abstract: Using cross-sectional data from the 2015 Canadian Community Health Survey–Nutrition, we aimed to identify and characterize the top 10 most frequently consumed plant-based foods and red/processed meat dishes in the Canadian population. Plant-based foods and red/processed meat dishes categories included 659 and 265 unique food codes, respectively, from the Canadian Nutrient File. A total of 20,176 Canadian individuals aged ≥1 year were included in our analysis. The most frequently consumed plant-based food was "Cooked regular long-grain white rice", which made a significant contribution to energy (12.1 ± 0.3%) and protein (6.1 ± 0.2%) intake among consumers. The most frequently consumed red/processed meat dish in Canada was "Cooked regular, lean or extra lean ground beef or patty". Among red/processed meat dishes, "ham and cheese sandwich with lettuce and spread" made the most significant contribution to the intake of energy (21.8 ± 0.7%), saturated fat (31.0 ± 1.0%), sodium (41.8 ± 1.3%), and sugars (8.2 ± 0.5%) among the consumers. Ground beef is the most frequently consumed red/processed meat dish and white rice is the most frequently consumed plant-based food among Canadians. Red/processed meat dishes are major drivers of the excessive intake of nutrients-to-limit.

Keywords: red/processed meat dishes; plant-based foods; long-grain white rice; nutrients to limit; Canadian population

Citation: Shafiee, M.; Islam, N.; Ramdath, D.D.; Vatanparast, H. Most Frequently Consumed Red/Processed Meat Dishes and Plant-Based Foods and Their Contribution to the Intake of Energy, Protein, and Nutrients-to-Limit among Canadians. *Nutrients* **2022**, *14*, 1257. https://doi.org/10.3390/nu14061257

Academic Editors: Colin Bell and Penny Love

Received: 16 February 2022
Accepted: 14 March 2022
Published: 16 March 2022

Publisher's Note: MDPI stays neutral with regard to jurisdictional claims in published maps and institutional affiliations.

1. Introduction

In 2017, dietary risk factors were responsible for 255 million disability-adjusted life-years (DALYs) and 11 million deaths across 195 countries [1]. Globally, the three major dietary risk factors for mortality and DALYs were high intake of sodium, low intake of whole grains, and low intake of fruits [1]. Health Canada recommends the regular intake of nutritious foods (e.g., whole grains, fruit, vegetables) that are commonly found in dietary patterns such as Dietary Approaches to Stop Hypertension (DASH) and Mediterranean-style diets, which are known to be associated with beneficial effects on human health [2–4]. According to Canada's new food guide (2019), consumption of nutritious foods often leads to low intakes of saturated fat (<10% of total energy intake), free sugars (<10% of total energy intake), or sodium (<2300 mg/day) [5]. While many animal-based foods are nutritious, Canada's new food guide (2019) emphasizes the consumption of more plant-based foods because of the positive health effects associated with higher intakes of vegetables and fruit, nuts, soy protein, and dietary fiber [5]. Moreover, dietary shifts toward fewer animal-based foods and more plant-based foods could encourage lower intakes of processed meat (such as sausages, ham, and hot dogs), and foods high in saturated fat [5].

In this regard, Kirkpatrick et al. reported that red meat (beef, pork, lamb, and goat) mixed dishes are the largest contributors to saturated fat and sodium intake in the diet of Canadians [6]. Further, it has been found that rice (84%) is the most commonly consumed plant-based food, and beef (48.7%) is the most commonly consumed animal-based food among the Brazilian population aged 10 years or over [7]. A growing body of evidence suggests that consumption of red and processed meat is not only associated with poorer health outcomes but also with negative environmental impacts [8–10]. According to the EAT-Lancet Commission report, a diet with fewer animal source foods and rich in plant-based foods confers environmental benefits [11]. In this regard, a study conducted in the European Union (EU) showed that replacing 25–50% of animal-derived foods (i.e., eggs, dairy products and meat) in the EU with plant-based foods on a dietary energy basis would lead to less per capita use of cropland for food production (23%), a reduction in greenhouse gas emissions (25–40%), and a reduction in nitrogen emissions (40%) [12]. Examining the most frequently consumed plant-based foods and red/processed meat dishes and their contribution to dietary components gives us an overview of the current dietary habits of Canadians and provides insights into targets for interventions to support healthy eating patterns.

The primary objective of the present study was to identify the top 10 most frequently consumed plant-based foods and red/processed meat dishes in the Canadian population aged ≥1 year in 2015. We also aimed to (1) rank the top 10 plant-based foods and red/processed meat dishes based on their contribution to nutrients-to-limit, energy, and protein, and (2) determine the mean and percentage contribution of the top five plant-based foods and red/processed meat dishes to nutrients-to-limit, energy, and protein intake of consumers.

2. Materials and Methods

2.1. Study Population and Dietary Data Collection

Cross-sectional data from the Canadian Community Health Survey (CCHS)–Nutrition 2015 was used in this study. These survey data were collected from 20,487 individuals across ten provinces of Canada, excluding individuals living in Indigenous settlements and reserves, Canadian forces employees, and the institutionalized population [13,14]. The survey respondents provided 24 h dietary recall information as well as sociodemographic and supplement use information. For this study, we used day 1 of the 24 h recall. The dietary recall data included all foods and beverages consumed by participants within a 24 h period as well as frequency, time, location, and amount of food consumed. These also included information on food groups, nutrients, and eating patterns. The Automated Multiple Pass Method (AMPM) was used to derive dietary information [14]. This method is based on the United States Department of Agriculture (USDA) and it is an automated questionnaire that maximizes the survey respondent's response to report and recall dietary intake in the last 24 h. A proxy interview was used to collect information from children aged 1–6 years under the supervision of parents or guardians. Children aged 6–11 years participated with their parental guidance, and respondents aged ≥12 years provided information using a non-proxy method. Detailed information on the survey design and methodology of the CCHS 2015–Nutrition can be found on the Statistics Canada website [13,14]. We accessed data at the Research Data Centre (RDC) of Statistics Canada. This study was exempt from ethics approval, since we used secondary data from a national survey conducted by Statistics Canada.

2.2. Analytical Sample

This study included all individuals aged ≥1 year, excluding pregnant and lactating women and individuals with no dietary data, resulting in a final sample size of 20,176. Individuals more than 1 year of age were added to the analyses because, from this age, children are introduced to solid food and breastfeeding is often discontinued.

2.3. Plant-Based Food and Red/Processed Meat Dishes Categories

The red/processed meat dishes that we included in this study were any beef/pork/lamb/goat dishes. This category included a range of 265 unique food codes from the Canadian Nutrient File (CNF). Any red/processed meats that are not part of any dishes were not included. Plant-based food categories included nuts, seeds and nuts, seed mixes, trail mixes, plant-based beverages, nut butters, legume dishes with meat, legume dishes without meat, Mexican dishes, rice and rice mixed dishes, soups, canned/jarred vegetables and legumes, legumes, grains, tofu and meat substitutes and plant-based non-dairy desserts. Fruits and vegetables were excluded from this study given their low consumption and minor contribution to protein intakes. This category included a range of 659 unique food codes from the CNF.

2.4. Statistical Analyses

We identified the top 10 most frequently consumed plant-based foods and red/processed meat dishes among the Canadian population. If an individual reported any amount of a particular food item (i.e., if serving size > 0 of the specific food code), then it was considered as one eating occasion of that specific food code. If any individual has the same food more than once a day, that person contributed to more than one eating occasion. Among the top 10 most frequently consumed plant-based foods and red/processed meat dishes, we ranked the highest contributors of energy, protein, saturated fat, sodium, and sugars to the daily intake of Canadians. We reported the percentage of individuals consuming the top five plant-based foods and red/processed meat dishes. Further, the mean amount and the mean proportion of energy, protein, saturated fat, sodium, and sugars derived from the top five plant-based foods and red/processed meat dishes were reported per consumer. For the mean percent contribution, individual percentages were calculated first and then the average was taken of that individual's percentages. All analyses were performed using SAS (version 9.3). We used appropriate bootstrapping weights to obtain population-level estimates using Statistics Canada guidelines [15]. The values were reported as mean $\pm$ SE or % $\pm$SE for continuous or categorical variables, respectively.

3. Results

Table 1 represents the top 10 most frequently reported plant-based foods and red/processed meat dishes among Canadians aged $\geq$1 year along with the percentages of individuals reporting the food. The most frequently consumed red/processed meat dish was "Cooked regular, lean or extra lean ground beef or patty". "Ham and cheese sandwich with lettuce and spread" and "ham sandwich with lettuce and spread" were the second and third most frequently consumed red/processed meat dishes. "Frankfurter (wiener/hot dog) on bun", "homemade-style spaghetti sauce", "barbecued pork", and "shepherd's pie" were also among the top 10 most frequently reported red/processed meat dishes. The most frequently consumed plant-based food in the Canadian population was "Cooked regular long-grain white rice", followed by "canned red ripe tomatoes". "Canned tomato puree, no salt added", and "smooth type peanut butter, fat, sugar and salt added" were the third and fourth most frequently consumed plant-based foods. "Canned cream of asparagus soup, ready-to-serve beef or chicken broth soup", "canned tomato sauce", "sweetened enriched almond milk", and "dried almonds" were also among the top 10 most frequently reported plant-based foods.

Ranking of the top 10 most frequently reported red/processed meat dishes based on their contribution to nutrients-to-limit, energy, and protein intake is presented in Table 2. Among the top 10 most frequently reported red/processed meat dishes, "ham and cheese sandwich with lettuce and spread" was the largest contributor to energy, protein, and saturated fat intake, and the second-largest contributor to sugars and sodium intake within the red/processed meat dishes category. Moreover, among red/processed meat dishes, "ham sandwich with lettuce and spread" was the top contributor to sodium intake, and "frankfurter (wiener/hot dog), with ketchup and/or mustard, on bun" was the top contributor to sugars intake in the Canadian diet within the red/processed meat dishes category.

Table 1. Top 10 most frequently reported plant-based foods and red/processed meat dishes among individuals ≥ 1 year in Canada (*n* = 20,176), 2015 Canadian Community Health Survey.

Rank	Red/Processed Meat Dishes	% of Individuals Reporting the Red/Processed Meat Dishes	Plant-Based Foods	% of Individuals Reporting the Plant-Based food
1	Ground beef or patty, cooked, NS as to regular, lean, or extra lean	4.6	Grains, rice, white, long-grain, regular, cooked	15.5
2	Sandwich, ham and cheese, with lettuce and spread	2.6	Tomato, red, ripe, canned, whole	13.4
3	Sandwich, ham, with lettuce and spread	2.9	Tomato products, canned, puree, no salt added	10.1
4	Frankfurter, wiener, or hot dog, with ketchup and/or mustard, on bun	2	Peanut butter, smooth type, fat, sugar and salt added	10.2
5	Frankfurter, wiener, or hot dog, plain, on bun	1.1	Soup, canned, cream of asparagus, condensed, milk added	7.5
6	Spaghetti sauce with beef or meat other than lamb or mutton, homemade-style	1.7	Soup, broth, beef, ready-to-serve	6.6
7	Spaghetti sauce with meat and vegetables, homemade-style	1.2	Tomato products, canned, sauce	5.9
8	Spaghetti sauce with a combination of meats, homemade-style	1.1	Soup, broth, chicken, ready-to-serve	4.8
9	Pork, spareribs, barbecued, with sauce, NS as to fat eaten	0.8	Plant-based beverage, almond, enriched, sweetened, vanilla flavored	2.8
10	Shepherd's pie, with corn	0.8	Nuts, almonds, dried, unblanched, unroasted	3.3

NS: Not specified.

Table 3 reports the ranking of the top 10 most frequently reported plant-based foods based on their contribution to energy, protein, saturated fat, sodium, and sugars intake. Among plant-based foods, "cooked regular long-grain white rice" was the largest contributor to energy and protein intake within the plant-based food category. In addition, "smooth type peanut butter with added fat, sugar and salt" was the top contributor to saturated fat intake, "ready-to-serve beef broth soup" was the top contributor to sodium intake, and "vanilla flavored sweetened enriched almond milk" was the top contributor to sugars intake among the top 10 most frequently reported plant-based foods within the plant-based food category.

Table 2. Ranking of the top 10 most frequently reported red/processed meat dishes based on their contribution to nutrients-to-limit, energy, and protein intake (*n* = 20,176), 2015 Canadian Community Health Survey.

	Red/Processed Meat Dishes	Dietary Components and Rank				
		Energy	Protein	SFA	Sugars	Sodium
1	Ground beef or patty, cooked, NS as to regular, lean, or extra lean	5	3	3	10	9
2	Sandwich, ham and cheese, with lettuce and spread	1	1	1	2	2
3	Sandwich, ham, with lettuce and spread	2	2	4	3	1
4	Frankfurter, wiener, or hot dog, with ketchup and/or mustard, on bun	3	4	2	1	3
5	Frankfurter, wiener, or hot dog, plain, on bun	6	9	6	9	7
6	Spaghetti sauce with beef or meat other than lamb or mutton, homemade-style	8	6	8	4	4
7	Spaghetti sauce with meat and vegetables, homemade-style	10	10	10	7	5
8	Spaghetti sauce with combination of meats, homemade-style	9	8	9	5	6
9	Pork, spareribs, barbecued, with sauce, NS as to fat eaten	7	7	5	6	10
10	Shepherd's pie, with corn	4	5	7	8	8

NS: Not specified; SFA: Saturated fatty acids.

Table 3. Ranking of the top 10 most frequently reported plant-based foods based on their contribution to nutrients-to-limit, energy, and protein intake (*n* = 20,176), 2015 Canadian Community Health Survey.

	Plant-Based Foods	Dietary Components and Rank				
		Energy	Protein	SFA	Sugars	Sodium
1	Grains, rice, white, long-grain, regular, cooked	1	1	4	8	8
2	Tomato, red, ripe, canned, whole	7	7	5	4	5
3	Tomato products, canned, puree, no salt added	6	6	7	3	7
4	Peanut butter, smooth type, fat, sugar and salt added	2	2	1	2	4
5	Soup, canned, cream of asparagus, condensed, milk added	3	5	2	9	9
6	Soup, broth, beef, ready-to-serve	9	4	8	6	1
7	Tomato products, canned, sauce	8	8	6	5	3
8	Soup, broth, chicken, ready-to-serve	10	9	9	10	2
9	Plant-based beverage, almond, enriched, sweetened, vanilla flavored	5	10	10	1	6
10	Nuts, almonds, dried, unblanched, unroasted	4	3	3	7	10

SFA: Saturated fatty acids.

Table 4 presents the proportion of individuals consuming the top five most frequently reported red/processed meat dishes as well as the mean and percentage contribution of each food item to energy, protein, saturated fat, sugars, and sodium intake of consumers on any given day. Since these results are based on 24 h recall, the proportion of individuals consuming the top five food sources are based on any given day. The proportion of Canadians consuming "cooked regular, lean or extra lean ground beef or patty" was 4.6%, followed by 2.9% for "ham sandwich with lettuce and spread". Among red/processed meat dishes, "ham and cheese sandwich with lettuce and spread" made the largest contribution to energy (21.8%), protein (31.4%), saturated fat (31.0%), and sodium (41.8%) intake among its consumers. In addition, "frankfurter (wiener/hot dog), with ketchup and/or mustard, on bun" accounted for 10.9% of the sugars intake of the consumers of this food item.

As reported in Table 5, the proportion of Canadians consuming "cooked regular long-grain white rice" was 15.5%, followed by 13.4% for "canned red ripe tomato". Among plant-based foods, "cooked regular long-grain white rice" made the largest contribution to energy intake (12.1%), followed by "smooth type peanut butter with added fat, sugar and salt" (8.4%). In addition, "smooth type peanut butter with added fat, sugar and salt" made the largest contribution to protein (8.1%), saturated fat (13.3%), sodium (4.8%), and sugars (3.8%) intake among its consumers.

Table 4. The mean and percentage contribution of the top five reported red/processed meat dishes to nutrients-to-limit, energy, and protein intake of Canadians aged ≥1 year in Canada, 2015 Canadian Community Health Survey.

Top 5 Red/Processed Meat Dishes	% Population (% ±SE)	Dietary Components [1]				
		Energy	Protein	SFA	Sugars	Sodium
# 1 Ground beef or patty, cooked, NS as to regular, lean, or extra lean	4.6 ± 0.4					
Daily intake (mean ± SE)		74.7 ± 3.57	7.9 ± 0.4	1.8 ± 0.1	0	116.2 ± 5.6
% Contribution per consumer/day (% ±SE)		4.0 ± 0.2	10.5 ± 0.5	9.5 ± 0.6	0	4.3 ± 0.2
# 2 Sandwich, ham and cheese, with lettuce and spread	2.6 ± 0.2					
Daily intake (mean ± SE)		421.5 ± 14.9	24.8 ± 0.9	7.8 ± 0.4	5.8 ± 0.2	1408.5 ± 54.6
% Contribution per consumer/day (% ±SE)		21.8 ± 0.7	31.4 ± 1.1	31.0 ± 1.0	8.2 ± 0.5	41.8 ± 1.3
# 3 Sandwich, ham, with lettuce and spread	2.9 ± 0.3					
Daily intake (mean ± SE)		332.9 ± 18.6	19.2 ± 1.0	3.3 ± 0.4	5.6 ± 0.3	1226.7 ± 59.9
% Contribution per consumer/day (% ±SE)		17.7 ± 0.8	25.2 ± 1.1	14.6 ± 1.0	7.1 ± 0.5	38.0 ± 1.3
# 4 Frankfurter, wiener, or hot dog, with ketchup and/or mustard, on bun	2.0 ± 0.2					
Daily intake (mean ± SE)		351.7 ± 19.2	14.0 ± 0.8	5.8 ± 0.3	8.3 ± 0.5	1154.5 ± 51.2
% Contribution per consumer/day (% ±SE)		18.1 ± 0.8	20.6 ± 1.1	24.0 ± 1.1	10.9 ± 0.9	36.9 ± 1.3
# 5 Frankfurter, wiener, or hot dog, plain, on bun	1.1 ± 0.2					
Daily intake (mean ± SE)		257.8 ± 15.9	10.9 ± 0.6	5.3 ± 0.4	3.9 ± 0.5	844.4 ± 50.8
% Contribution per consumer/day (% ±SE)		13.7 ± 0.8	16.4 ± 1.2	22.4 ± 0.2	6.5 ± 1.5	27.6 ± 1.3

NS: Not specified; SFA: Saturated fatty acids. [1] Units: Energy in kcal, Protein in grams, SFA in grams, Sugars in grams, and Sodium in mg.

Table 5. The mean and percentage contribution of the top five reported plant-based foods to nutrients-to-limit, energy, and protein intake of Canadians aged ≥1 year in Canada, 2015 Canadian Community Health Survey.

Top 5 Plant-Based Foods	% Population (% ±SE)	Dietary Components [1]				
		Energy	Protein	SFA	Sugars	Sodium
# 1 Grains, rice, white, long-grain, regular, cooked	15.5 ± 0.6					
Daily intake (mean ± SE)		199.4 ± 5.8	4.1 ± 0.1	0.1 ± 0.0	0.1 ± 0.0	1.5 ± 0.04
% Contribution per consumer/day (% ±SE)		12.1 ± 0.3	6.1 ± 0.2	0.9 ± 0.04	0.2 ± 0.0	0.09 ± 0.0
# 2 Tomato, red, ripe, canned, whole	13.4 ± 0.5					
Daily intake (mean ± SE)		7.6 ± 0.3	0.4 ± 0.01	0.02 ± 0.0	1.2 ± 0.1	54.3 ± 2.2
% Contribution per consumer/day (%)		0.3 ± 0.01	0.5 ± 0.02	0.07 ± 0.0	1.7 ± 0.1	1.9 ± 0.07
# 3 Tomato products, canned, puree, no salt added	10.1 ± 0.4					
Daily intake (mean ± SE)		14.6 ± 0.7	0.6 ± 0.03	0.01 ± 0.0	1.9 ± 0.09	10.7 ± 0.5
% Contribution per consumer/day (% ±SE)		0.7 ± 0.03	0.9 ± 0.03	0.05 ± 0.0	0.03 ± 0.01	0.3 ± 0.01
# 4 Peanut butter, smooth type, fat, sugar and salt added	10.2 ± 0.5					
Daily intake (mean ± SE)		168.1 ± 8.5	6.2 ± 0.3	2.8 ± 0.2	2.9 ± 0.2	119.8 ± 6.1
% Contribution per consumer/day (% ±SE)		8.4 ± 0.03	8.1 ± 0.3	13.3 ± 0.05	3.8 ± 0.2	4.8 ±0.2
# 5 Soup, canned, cream of asparagus, condensed, milk added	7.5 ± 0.4					
Daily intake (mean ± SE)		45.4 ± 3.1	1.1 ± 0.1	0.4 ± 0.03	0.2 ± 0.01	0.2 ± 0.01
% Contribution per consumer/day (% ±SE)		2.5 ± 0.2	1.6 ± 0.1	2.3 ± 0.2	0.3 ± 0.03	0.01 ± 0.0

SFA: Saturated fatty acids. [1] Units: Energy in kcal, Protein in grams, SFA in grams, Sugars in grams, and Sodium in mg.

4. Discussion

This is the first study to identify the top 10 most frequently consumed plant-based foods and red/processed meat dishes in the Canadian population, using a nationally representative sample. Overall, the most frequently consumed plant-based food in Canada was "cooked regular long-grain white rice". The mean contribution of this food item to the energy intake of its consumers was just under 200 kcal/day. The most frequently consumed red/processed meat dish was "cooked regular, lean or extra lean ground beef or patty". Among all red/processed meat dishes, "ham and cheese sandwich with lettuce and spread" made the largest contribution to energy, protein, saturated fat, and sodium intake.

Rice is the most widely consumed food staple for almost 50% of the world's population [16]. On a global basis, rice provides 21% of energy and 15% of protein per capita [17], and long-grain white rice is known to be the most commonly eaten form of rice [18]. Similarly, our results revealed that the most frequently consumed plant-based food in Canada in 2015 was "cooked regular long-grain white rice". More than 15% of Canadians reported consuming this food item, and it contributed to 12.1% of energy intake and 6.1% of protein intake among consumers. Using data from the CCHS 2015, Kirkpatrick et al. reported that rice and rice mixed dishes were among the top 10 contributors to energy intake in the general Canadian population (≥1 year), and among the top five contributors to energy intake in the low-income group [6]. In a study investigating the dietary sources of energy and nutrient intakes among five ethnic groups (i.e., Caucasian; Latino; Native Hawaiian; Japanese-American; African American) in the U.S., it was found that rice made a significant contribution to dietary energy intake, ranging from 5.3% (Caucasian women) to 22.9% (Japanese-American men). In addition, rice was found to be the top dietary source of protein for Japanese-American men and women, respectively, contributing 12.7% and 10.4% to protein intake [19]. Sharma et al. also reported that white rice was the most commonly consumed grain food among Japanese-American, Native Hawaiian, and Caucasian men (12.0–44.1%), and the most commonly consumed refined grain among all ethnic groups (10.3–54.1%), except for Latinos [20]. In another study aiming to describe the most commonly consumed foods in the Brazilian diet, Souza et al. found that rice (84%) was the most frequently recorded food by Brazilian individuals (≥10 years), followed by coffee (79%) and beans (72.8%) [7]. Using data from the 2009–2013 Korea National Health and Nutrition Examination Surveys (KNHANES), it was revealed that white rice was the major source of energy (31%) among Korean preschoolers aged 1–5 years, followed by milk (10.2%) and bread (3.5%) [21]. Although white rice (milled and/or polished rice) is the most commonly consumed form of rice, it is a poor source of vitamins and minerals due to removing the bran and germ layers of the seed during the milling process [22]. In their review of the literature, Saleh et al. reported that brown rice is nutritionally superior to white rice because of higher levels of nutrients such as protein, vitamins, and minerals [22]. In accordance with this, the new Canada's Food Guide has placed a major emphasis on whole grain foods and recommended regular consumption of whole grains and decreasing consumption of refined grains [5].

Our results showed that the most frequently consumed red/processed meat dish in Canada in 2015 was "cooked regular, lean or extra lean ground beef or patty". This food item was consumed by 4.6% of the Canadian population and accounted for 4% of energy intake, 9.5% of saturated fat intake, and 10.5% of protein intake among its consumers. Ground beef is the most commonly consumed form of beef in the U.S., accounting for more than 40% of all beef consumed [23]. Results from a multiethnic cohort study of 215,000 individuals aged 45–75 years showed that lean beef (steak/roast) was the most commonly consumed red meat for all ethnic groups in the U.S. (9.3–14.3%), except for Japanese-Americans and Native Hawaiians [24]. Using data from the 2003–2006 National Health and Nutrition Examination Survey (NHANES), O'Neil et al. reported that beef was the top source of protein (14.0%), the third-highest ranked source of saturated fat (7.9%), and the fourth-highest ranked source of energy (5.0%) among American adults aged ≥19 years [25]. Using the same survey data, Huth et al. found that the top three food sources of saturated fat in the diet of Americans aged ≥2 years were cheese (16.5%), beef (8.5%), and milk (8.3%) [26]. In addition, results from the 2007–2010 NHANES showed that ground beef contributed to 5.6% of animal protein intake and 2.6% of total protein intake among U.S. adults aged 19 years and older [27]. Souza et al. reported that beef was the fifth most commonly reported food item, and the most frequently reported animal-based food on the first day's food record of Brazilian individuals (≥10 years) [7]. A more recent study also showed that beef was the most commonly consumed meat among Brazilian individuals aged 10 years and older (49%), and the mean beef intake for the entire country was 63 g/day [28]. In a survey conducted in Spanish adults aged 25 to 75 years, beef was found to be the most frequently consumed red meat (63.6%), with pork in second place (52.6%) [29]. Using data from the 2011–2012 National Nutrition and Physical Activity Survey (NNPAS), Sui et al.

reported that red meat was consumed by 48.6% of Australian men and women, with beef as the most frequently reported type (41.8% and 34.7%, respectively) [30]. Some observational studies have found an association between consumption of beef and increased risk of a number of cancers [31–34]. Therefore, reducing beef intake and replacing it with plant-based proteins could be an effective strategy in the prevention of such conditions. In line with this, the new Canada's Food Guide has recommended regular consumption of protein foods, with particular emphasis on plant-based proteins [5]. Further, according to the new food guide, ruminant animal-based foods such as beef and lamb are natural sources of trans fat, a type of unsaturated fat known to have adverse health effects [5].

Processed meats, such as ham, bacon, frankfurters, and sausages that have been modified through salting, curing, fermentation, or smoking, account for a large proportion of the world's meat consumption [35]. Our results revealed that "ham and cheese sandwich with lettuce and spread" was the second most frequently consumed red/processed meat dish in Canada in 2015. This food item accounted for 41.8% of sodium intake, 31.4% of protein intake, 31.0% of saturated fat intake, and 21.8% of energy intake among its consumers. Among red/processed meat dishes, "ham sandwich with lettuce and spread" and "frankfurter (wiener/hot dog) with ketchup and/or mustard on bun" were in the next rankings. Using data from the 2009–2010 NHANES, Sebastian et al. reported that nearly half of American adults (49%) reported eating a sandwich on any given day, and the mean contribution of sandwiches to energy intakes of all adults was 200 kcal for women and 350 kcal for men [36]. In addition, the mean contribution of sandwiches to sodium intake of American men and women was 902 mg and 489 mg, respectively [36]. In the U.S., processed meat intake constitutes 22% of the total meat consumed from either poultry or red meat categories [37]. Using data of Australians aged ≥2 years from the 2011–2012 NNPAS, Sui et al. found that 37.8% of participants reported consuming processed meat, and ham was the most frequently reported type of processed meat (females 16.8%, males 19.4%), followed by bacon (females 12.4%, males 15.3%), and sausage (females 5.8%, males 8.5%) [30]. Kirkpatrick et al. reported that processed meats are among the top 10 Contributors to sodium and saturated fat intake of Canadians aged 1 year and older [6]. In a study aiming to identify the major food sources of dietary sodium using 3-day food records, it was found that processed meat was the largest contributor to daily sodium intake among Mexican adults, representing 8% of total sodium intake per capita [38]. The results also showed that total sodium contributed by processed meat in the entire sample population was 223 mg/day [38]. Using data obtained from 21,108 British households, Ni Mhurchu et al. observed that processed meat (18%) was the second largest contributor to sodium purchases after table salt (23%) [39]. In a nationwide survey conducted in Brazil, it was found that processed meat was among the top five contributors to saturated fat intake among Brazilian individuals aged ≥10 years [40]. Moreover, increased morbidity and mortality related to high consumption of processed meat has been linked to their high content of sodium and saturated fat [41–44]. We have recently shown that reducing red and processed meat by half and increasing plant-based meat alternatives by 100% may assist in reducing the intake of sodium and saturated fat, and increase the overall nutritional value of the diet [45].

Strengths and Limitations

This study used nutrition data from CCHS–Nutrition 2015, a nationally representative survey of the Canadian population aged one year and older. A major strength of this study was the opportunity to identify the top 10 most frequently consumed plant-based foods and red/processed meat dishes, shortly after the introduction of the new Canada's Food Guide, which places a major emphasis on the consumption of plant-based foods. We also acknowledge some limitations. First, since we combined age and sex groups in this study, some results may vary for different age/sex groups. Second, detailed dietary data were obtained using a 24 h dietary recall and, therefore, may not reflect the usual intake of the participants. In addition, the 24 h dietary recall is a self-report method subject to recall bias and misreporting (i.e., overestimating or underestimating dietary intake). Third,

CCHS–Nutrition 2015 does not include added sugar intake information of Canadians. That is why we were only able to report the total sugar intake for the analyses. Fourth, the descriptive, cross-sectional design of the study does not allow us to determine the causal relationship between patterns of food consumption and health outcomes. Fifth, we included the consumption data of only two main food categories, namely plant-based foods and red/processed meat dishes. It is notable that the plant-based foods make up a broader category compared to the red/processed meat dishes category, and include vegetables and fruit, grain products, and protein-based foods.

5. Conclusions

The results of this study revealed that the most frequently consumed plant-based food among the Canadian population aged ≥1 year was "cooked regular long-grain white rice", followed by "canned red ripe tomatoes". Cooked regular long-grain white rice made a significant contribution to energy and protein intake among the consumers of this food item. Among the top five most frequently consumed plant-based foods, "smooth type peanut butter, fat, sugar, and salt added" made the most significant contribution to saturated fat, protein, sugars, and sodium intake. The most frequently consumed red/processed meat dish in Canada was "Cooked regular, lean or extra lean ground beef or patty". Further, the top five red/processed meat dishes, especially "ham and cheese sandwich with lettuce and spread", made a significant contribution to the intake of saturated fat, sugars, and sodium in the diet of Canadians. According to the new Canada's Food Guide, patterns of eating that place a major emphasis on plant-based foods typically result in higher intakes of dietary fiber, nuts, vegetables and fruit, and soy protein, and also encourage lower intakes of processed meat and saturated fat [5]. Thus, minimizing the consumption of red/processed meat dishes and shifting intakes towards more plant-based foods may improve the health of Canadians and confer environmental benefits [5,11]. Putting this into practice, in the most recent Canada's Food Guide, Health Canada has merged the two food groups (Meat & Alternatives and Milk & Alternatives) of the 2007 Food Guide into a single group called "Protein Foods" [5]. Among protein foods, health Canada also recommended consuming plant-based more often. The new guidelines along with relevant health promotion initiatives may have an impact on shifting toward consumption of more plant-based foods among Canadians. Further research is required to determine and compare the most frequently consumed red/processed meat dishes and plant-based foods among different age/sex groups of Canadians and how they can contribute to the intake of energy, protein, and nutrients-to-limit.

Author Contributions: Conceptualization, H.V. and D.D.R.; Formal analysis, H.V. and N.I.; Funding acquisition, H.V. and D.D.R.; Investigation, H.V.; Methodology, H.V., N.I. and M.S.; Project administration, H.V.; Resources, H.V.; Supervision, H.V.; Writing—original draft, H.V., N.I. and M.S.; Writing—review and editing, H.V., N.I., M.S. and D.D.R. All authors have read and agreed to the published version of the manuscript.

Funding: This work was supported by a grant provided by Saskatchewan Pulse Growers Association (funding number: 350778)

Institutional Review Board Statement: Ethical review and approval were waived for this study due to using secondary data.

Informed Consent Statement: Informed consent was obtained from all subjects involved in the study.

Data Availability Statement: The data will be available at the Research Data Centre (RDC) of Statistics Canada.

Acknowledgments: The analysis presented in this paper was conducted at the Saskatchewan RDC, which is part of the Canadian Research Data Centre Network (CRDCN). We thank Ruben Mercado, Analyst, and Saskatchewan RDC for all his cooperation in the data access and vetting process. We also thank Anne Kennedy and Sheryl Conrad of Agriculture and Agri-Food Canada for assisting in conceptualization of the study.

Conflicts of Interest: The authors declare no conflict of interest.

References

1. Afshin, A.; Sur, P.J.; Fay, K.A.; Cornaby, L.; Ferrara, G.; Salama, J.S.; Mullany, E.C.; Abate, K.H.; Abbafati, C.; Abebe, Z. Health effects of dietary risks in 195 countries, 1990–2017: A systematic analysis for the Global Burden of Disease Study 2017. *Lancet* **2019**, *393*, 1958–1972. [CrossRef]
2. Saneei, P.; Salehi-Abargouei, A.; Esmaillzadeh, A.; Azadbakht, L. Influence of Dietary Approaches to Stop Hypertension (DASH) diet on blood pressure: A systematic review and meta-analysis on randomized controlled trials. *Nutr. Metab. Cardiovasc. Dis.* **2014**, *24*, 1253–1261. [CrossRef]
3. Garcia, M.; Bihuniak, J.D.; Shook, J.; Kenny, A.; Kerstetter, J.; Huedo-Medina, T.B. The effect of the traditional Mediterranean-style diet on metabolic risk factors: A meta-analysis. *Nutrients* **2016**, *8*, 168. [CrossRef]
4. Dinu, M.; Pagliai, G.; Casini, A.; Sofi, F. Mediterranean diet and multiple health outcomes: An umbrella review of meta-analyses of observational studies and randomised trials. *Eur. J. Clin. Nutr.* **2018**, *72*, 30–43. [CrossRef]
5. Health Canada. Canada's Food Guide. Available online: https://food-guide.canada.ca/en/ (accessed on 13 July 2021).
6. Kirkpatrick, S.I.; Raffoul, A.; Lee, K.M.; Jones, A.C. Top dietary sources of energy, sodium, sugars, and saturated fats among Canadians: Insights from the 2015 Canadian Community Health Survey. *Appl. Physiol. Nutr. Metab. Physiol. Appl. Nutr. et Metab.* **2019**, *44*, 650–658. [CrossRef]
7. Souza, A.D.M.; Pereira, R.A.; Yokoo, E.M.; Levy, R.B.; Sichieri, R. Most consumed foods in Brazil: National dietary survey 2008–2009. *Rev. Saude Publica* **2013**, *47*, 190s–199s. [CrossRef]
8. Wang, X.; Lin, X.; Ouyang, Y.Y.; Liu, J.; Zhao, G.; Pan, A.; Hu, F.B. Red and processed meat consumption and mortality: Dose–response meta-analysis of prospective cohort studies. *Public Health Nutr.* **2016**, *19*, 893–905. [CrossRef]
9. Clonan, A.; Wilson, P.; Swift, J.A.; Leibovici, D.G.; Holdsworth, M. Red and processed meat consumption and purchasing behaviours and attitudes: Impacts for human health, animal welfare and environmental sustainability. *Public Health Nutr.* **2015**, *18*, 2446–2456. [CrossRef]
10. Potter, J.D. Red and Processed Meat, and Human and Planetary Health. *BMJ.* **2017**, *357*, 9–10. [CrossRef]
11. Willett, W.; Rockström, J.; Loken, B.; Springmann, M.; Lang, T.; Vermeulen, S.; Garnett, T.; Tilman, D.; DeClerck, F.; Wood, A. Food in the Anthropocene: The EAT–Lancet Commission on healthy diets from sustainable food systems. *Lancet* **2019**, *393*, 447–492. [CrossRef]
12. Westhoek, H.; Lesschen, J.P.; Rood, T.; Wagner, S.; De Marco, A.; Murphy-Bokern, D.; Leip, A.; van Grinsven, H.; Sutton, M.A.; Oenema, O. Food choices, health and environment: Effects of cutting Europe's meat and dairy intake. *Glob. Environ. Chang.* **2014**, *26*, 196–205. [CrossRef]
13. Health Canada. Reference Guide to Understanding and Using the Data. Available online: Https://www.canada.ca/en/health-canada/services/food-nutrition/food-nutrition-surveillance/health-nutrition-surveys/canadian-community-health-survey-cchs/reference-guide-understanding-using-data-2015.html (accessed on 13 July 2021).
14. Statistics Canada. Canadian Community Health Survey—Nutrition (CCHS). Available online: Https://www23.statcan.gc.ca/imdb/p2SV.pl?Function=getSurvey&SDDS=5049 (accessed on 13 July 2021).
15. Statistics Canada. The Research Data Centres Information and Technical Bulletin-Weighted Estimation and Bootstrap Variance Estimation for Analyzing Survey Data: How to Implement in Selected Software. Available online: https://www150.statcan.gc.ca/n1/en/catalogue/12-002-X201400111901 (accessed on 13 July 2021).
16. Fitzgerald, M.A.; McCouch, S.R.; Hall, R.D. Not just a grain of rice: The quest for quality. *Trends Plant Sci.* **2009**, *14*, 133–139. [CrossRef]
17. Maclean, J.L.; Dawe, D.C.; Hettel, G.P.; Hardy, B. *Rice Almanac: Source Book for the Most Important Economic Activity on Earth*; International Rice Research Institute, CABI Publishing: Wallingford, UK, 2002.
18. Hoogenkamp, H.; Kumagai, H.; Wanasundara, J. Rice protein and rice protein products. In *Sustainable Protein Sources*; Academic Press: San Diego, CA, USA, 2017; pp. 47–65.
19. Sharma, S.; Wilkens, L.R.; Shen, L.; Kolonel, L.N. Dietary sources of five nutrients in ethnic groups represented in the Multiethnic Cohort. *Br. J. Nutr.* **2013**, *109*, 1479–1489. [CrossRef]
20. Sharma, S.; Sheehy, T.; Kolonel, L.N. Ethnic differences in grains consumption and their contribution to intake of B-vitamins: Results of the Multiethnic Cohort Study. *Nutr. J.* **2013**, *12*, 65. [CrossRef]
21. Kang, M.; Shim, J.E.; Kwon, K.; Song, S. Contribution of foods to absolute nutrient intake and between-person variations of nutrient intake in Korean preschoolers. *Nutr. Res. Pract.* **2019**, *13*, 323–332. [CrossRef]
22. Saleh, A.S.; Wang, P.; Wang, N.; Yang, L.; Xiao, Z. Brown rice versus white rice: Nutritional quality, potential health benefits, development of food products, and preservation technologies. *Compr. Rev. Food Sci. Food Saf.* **2019**, *18*, 1070–1096. [CrossRef]
23. Davis, C.G.; Lin, B.-H. *Factors Affecting US Beef Consumption*; US Department of Agriculture, Economic Research Service: Washington, DC, USA, 2005.
24. Sharma, S.; Sheehy, T.; Kolonel, L.N. Contribution of meat to vitamin B 12, iron and zinc intakes in five ethnic groups in the USA: Implications for developing food-based dietary guidelines. *J. Hum. Nutr. Diet.* **2013**, *26*, 156–168. [CrossRef]
25. O'Neil, C.E.; Keast, D.R.; Fulgoni, V.L.; Nicklas, T.A. Food sources of energy and nutrients among adults in the US: NHANES 2003–2006. *Nutrients* **2012**, *4*, 2097–2120. [CrossRef]

26. Huth, P.J.; Fulgoni, V.L.; Keast, D.R.; Park, K.; Auestad, N. Major food sources of calories, added sugars, and saturated fat and their contribution to essential nutrient intakes in the U.S. diet: Data from the National Health and Nutrition Examination Survey (2003–2006). *Nutr. J.* **2013**, *12*, 116. [CrossRef]
27. Pasiakos, S.M.; Agarwal, S.; Lieberman, H.R.; Fulgoni, V.L., III. Sources and Amounts of Animal, Dairy, and Plant Protein Intake of US Adults in 2007–2010. *Nutrients* **2015**, *7*, 7058–7069. [CrossRef]
28. Carvalho, A.M.; Selem, S.S.; Miranda, A.M.; Marchioni, D.M. Excessive red and processed meat intake: Relations with health and environment in Brazil. *Br. J. Nutr.* **2016**, *115*, 2011–2016. [CrossRef]
29. Escriba-Perez, C.; Baviera-Puig, A.; Buitrago-Vera, J.; Montero-Vicente, L. Consumer profile analysis for different types of meat in Spain. *Meat Sci.* **2017**, *129*, 120–126. [CrossRef]
30. Sui, Z.; Raubenheimer, D.; Rangan, A. Consumption patterns of meat, poultry, and fish after disaggregation of mixed dishes: Secondary analysis of the Australian National Nutrition and Physical Activity Survey 2011–12. *BMC Nutr.* **2017**, *3*, 52. [CrossRef]
31. Hamada, G.S.; Kowalski, L.P.; Nishimoto, I.N.; Rodrigues, J.J.G.; Iriya, K.; Sasazuki, S.; Hanaoka, T.; Tsugane, S. Risk factors for stomach cancer in Brazil (II): A case-control study among Japanese Brazilians in Sao Paulo. *Jpn. J. Clin. Oncol.* **2002**, *32*, 284–290. [CrossRef]
32. Figg, W.D. How do you want your steak prepared? The impact of meat consumption and preparation on prostate cancer. *Cancer Biol. Ther.* **2012**, *13*, 1141–1142. [CrossRef]
33. Hermans, K.E.; van den Brandt, P.A.; Loef, C.; Jansen, R.L.; Schouten, L.J. Meat consumption and cancer of unknown primary (CUP) risk: Results from The Netherlands cohort study on diet and cancer. *Eur. J. Nutr.* **2021**, *60*, 4579–4593. [CrossRef]
34. Deoula, M.S.; El Kinany, K.; Huybrechts, I.; Gunter, M.J.; Hatime, Z.; Boudouaya, H.A.; Benslimane, A.; Nejjari, C.; El Abkari, M.; Badre, W. Consumption of meat, traditional and modern processed meat and colorectal cancer risk among the Moroccan population: A large-scale case–control study. *Int. J. Cancer* **2020**, *146*, 1333–1345. [CrossRef]
35. Hung, Y.; de Kok, T.M.; Verbeke, W. Consumer attitude and purchase intention towards processed meat products with natural compounds and a reduced level of nitrite. *Meat Sci.* **2016**, *121*, 119–126. [CrossRef]
36. Sebastian, R.S.; Wilkinson Enns, C.; Goldman, J.D.; Hoy, M.K.; Moshfegh, A.J. Sandwiches are major contributors of sodium in the diets of American adults: Results from What We Eat in America, National Health and Nutrition Examination Survey 2009–2010. *J. Acad. Nutr. Diet.* **2015**, *115*, 272–277. [CrossRef]
37. Daniel, C.R.; Cross, A.J.; Koebnick, C.; Sinha, R. Trends in meat consumption in the USA. *Public Health Nutr.* **2011**, *14*, 575–583. [CrossRef]
38. Colin-Ramirez, E.; Espinosa-Cuevas, Á.; Miranda-Alatriste, P.V.; Tovar-Villegas, V.I.; Arcand, J.; Correa-Rotter, R. Food sources of sodium intake in an adult mexican population: A sub-analysis of the SALMEX study. *Nutrients* **2017**, *9*, 810. [CrossRef] [PubMed]
39. Ni Mhurchu, C.; Capelin, C.; Dunford, E.K.; Webster, J.L.; Neal, B.C.; Jebb, S.A. Sodium content of processed foods in the United Kingdom: Analysis of 44,000 foods purchased by 21,000 households. *Am. J. Clin. Nutr.* **2011**, *93*, 594–600. [CrossRef] [PubMed]
40. Pereira, R.A.; Duffey, K.J.; Sichieri, R.; Popkin, B.M. Sources of excessive saturated fat, trans fat and sugar consumption in Brazil: An analysis of the first Brazilian nationwide individual dietary survey. *Public Health Nutr.* **2014**, *17*, 113–121. [CrossRef] [PubMed]
41. Micha, R.; Michas, G.; Mozaffarian, D. Unprocessed red and processed meats and risk of coronary artery disease and type 2 diabetes—An updated review of the evidence. *Curr. Atheroscler. Rep.* **2012**, *14*, 515–524. [CrossRef] [PubMed]
42. Mozaffarian, D.; Micha, R.; Wallace, S. Effects on coronary heart disease of increasing polyunsaturated fat in place of saturated fat: A systematic review and meta-analysis of randomized controlled trials. *PLoS Med.* **2010**, *7*, e1000252. [CrossRef] [PubMed]
43. Kim, H.; Hu, E.A.; Rebholz, C.M. Ultra-processed food intake and mortality in the USA: Results from the Third National Health and Nutrition Examination Survey (NHANES III, 1988–1994). *Public Health Nutr.* **2019**, *22*, 1777–1785. [CrossRef] [PubMed]
44. Araya, C.; Corvalán, C.; Cediel, G.; Taillie, L.; Reyes, M. Ultra-processed food consumption among chilean preschoolers is associated with diets promoting non-communicable diseases. *Front. Nutr.* **2021**, *8*, 127. [CrossRef]
45. Vatanparast, H.; Islam, N.; Shafiee, M.; Ramdath, D.D. Increasing Plant-Based Meat Alternatives and Decreasing Red and Processed Meat in the Diet Differentially Affect the Diet Quality and Nutrient Intakes of Canadians. *Nutrients* **2020**, *12*, 2034. [CrossRef]

Review

Plant-Based Dietary Patterns for Human and Planetary Health

Joshua Gibbs [1,*] and Francesco P. Cappuccio [1,2,*]

1   WHO Collaborating Centre for Nutrition, Division of Health Sciences, Warwick Medical School, University of Warwick, Coventry CV4 7LA, UK
2   Department of Medicine, University Hospitals Coventry & Warwickshire NHS Trust, Coventry CV2 2DX, UK
*   Correspondence: joshua.gibbs@warwick.ac.uk (J.G.); f.p.cappuccio@warwick.ac.uk (F.P.C.)

Abstract: The coronavirus pandemic has acted as a reset on global economies, providing us with the opportunity to build back greener and ensure global warming does not surpass 1.5 °C. It is time for developed nations to commit to red meat reduction targets and shift to plant-based dietary patterns. Transitioning to plant-based diets (PBDs) has the potential to reduce diet-related land use by 76%, diet-related greenhouse gas emissions by 49%, eutrophication by 49%, and green and blue water use by 21% and 14%, respectively, whilst garnering substantial health co-benefits. An extensive body of data from prospective cohort studies and controlled trials supports the implementation of PBDs for obesity and chronic disease prevention. The consumption of diets high in fruits, vegetables, legumes, whole grains, nuts, fish, and unsaturated vegetable oils, and low in animal products, refined grains, and added sugars are associated with a lower risk of all-cause mortality. Meat appreciation, health concerns, convenience, and expense are prominent barriers to PBDs. Strategic policy action is required to overcome these barriers and promote the implementation of healthy and sustainable PBDs.

Keywords: plant-based diet; planetary health; human health; sustainability; chronic disease prevention

Citation: Gibbs, J.; Cappuccio, F.P. Plant-Based Dietary Patterns for Human and Planetary Health. *Nutrients* **2022**, *14*, 1614. https://doi.org/10.3390/nu14081614

Academic Editors: Colin Bell and Penny Love

Received: 9 March 2022
Accepted: 11 April 2022
Published: 13 April 2022

Publisher's Note: MDPI stays neutral with regard to jurisdictional claims in published maps and institutional affiliations.

1. Introduction

There is scientific consensus that anthropogenic greenhouse gas (GHG) emissions influence global warming and climate change [1]. To limit the negative consequences of climate change, 196 parties have committed to keep the increase in global average temperature below 2 °C above pre-industrial levels and try to limit warming to 1.5 °C [2]. The coronavirus pandemic has acted as a reset on global economies providing us with the opportunity to build back greener and maximize our chances of meeting the 1.5 °C target [3]. For example, the government of the United Kingdom (UK) has laid out a ten-point plan for a green industrial revolution in which they commit to transforming the energy sector, ending the sale of petrol and diesel cars, decarbonising public transport, developing greener buildings, investing in carbon capture and storage, and protecting the natural environment [4]. Worryingly, they failed to address agriculture in their plans. Revolutionizing agricultural systems should arguably be a top priority considering food production is the single largest cause of global environmental change [5]. Current agricultural practices constitute up to 30% of global anthropogenic GHG emissions [6] and 70% of freshwater use [7], whilst occupying approximately 40% of Earth's land [8]. Therefore, innovation within the agricultural sector has the potential to generate substantial sustainability gains.

A possible line of action, that is receiving ever-increasing interest, is to transition towards a plant-based food system. Plant-based foods have a significantly smaller footprint on the environment than animal-based foods. Even the least sustainable vegetables and cereals cause less environmental harm than the lowest impact meat and dairy products [9]. On top of the low environmental impact of plant-based diets (PBDs), they may provide additional benefits to human health. Unhealthy diets now represent the largest burden of disease globally, presenting a greater risk to morbidity, disability, and mortality than unsafe sex, alcohol, drug, and tobacco use combined [5]. Adopting plant-based food systems may allow countries to reduce their environmental footprints and tackle their obesity and diet-related non-communicable disease burdens simultaneously. A few reviews have covered the planetary and human health benefits associated with PBDs; however, since their publication, additional data of relevance have become available [10–12]. The aim of this review is to provide a concise summary of the planetary and human health benefits associated with PBDs using evidence from the latest advances in the field. This review will also summarise the main barriers to PBDs and offer potential solutions.

PBD is an umbrella term that describes any dietary pattern that emphasises the consumption of foods derived from plants and excludes or limits the consumption of most or all animal products. PBDs can be healthy or unhealthy depending on their composition. Healthy PBDs focus on unprocessed plant foods, including fruits, vegetables, whole grains, legumes, nuts, and seeds, whereas unhealthy PBDs contain high quantities of processed and ultra-processed plant foods such as sugar-sweetened beverages, refined grains, sweets, and desserts. Descriptions of the various PBDs mentioned in this review are shown in Table 1.

Table 1. Descriptions of various plant-based dietary patterns.

Dietary Pattern	Description
Healthful plant-based	High consumption of fruits, vegetables, legumes, whole grains, nuts, and unsaturated vegetable oils, and lower or no consumption of animal products (meat, fish, poultry, dairy, and eggs) and processed foods
Unhealthful plant-based	High consumption of fruit juices, sugar-sweetened beverages, refined grains, potatoes, and sweets and desserts, and lower consumption of animal products (meat, fish, poultry, dairy, and eggs) and healthy plant foods (fruits, vegetables, legumes, whole grains, nuts, and unsaturated vegetable oils).
Vegan	Excludes all animal products (meat, fish, poultry, dairy, and eggs) and is based solely on plant-based foods
Vegetarian	Excludes meat, fish, and poultry but does include eggs and dairy, in addition to plant-based foods
Pescatarian	Excludes meat and poultry but includes fish, dairy, and eggs, in addition to plant-based foods
Semi-vegetarian	Includes all animal products, including meat, fish, poultry, dairy, and eggs, in addition to plant-based foods. However, red meat intake is limited
EAT-Lancet reference	Consists of fruits and vegetables, whole grains, legumes, nuts, and unsaturated oils; low to moderate consumption of seafood and poultry; zero to low consumption of red meat, processed meat, added sugar, refined grains, and starchy vegetables

2. Planetary Health

2.1. Greenhouse Gas (GHG) Emission

Food systems are responsible for 21–37% of all GHG emissions globally [13]. Innovation and transformation within the food and agricultural sectors are imperative to limiting global warming to 1.5 °C. Between 2017 and 2018, agricultural emissions rose by 1.5% reaching a total of 5.6 $GtCO_2$, even with modest improvements in efficiency [14]. Of this total, 52% was caused by cattle products, primarily meat and dairy. Per-capita emissions from food consumption are 39% and 41% higher in very high human development index (HDI) countries than in high HDI countries and low HDI countries, respectively [14]. These differences in emissions are despite the use of high emission-intensity beef farming in low HDI countries. In very high HDI countries, cattle products are responsible for 68% of total consumption-based agricultural GHG emissions [14]. Reducing red meat consumption is a major key to meeting emission targets for very high HDI countries and it would deliver substantial health co-benefits. The rate of red meat-related mortality is nearly nine times greater in very high HDI countries than in low HDI countries [14]. Life cycle assessment studies have shown that pork, chicken, and seafood produce less GHG emissions than beef; however, even the lowest impact animal products exceed the average GHG emissions of substitute plant proteins [9,15]. Moving to diets that exclude animal products could reduce global GHG emissions by 49% (Figure 1) [9].

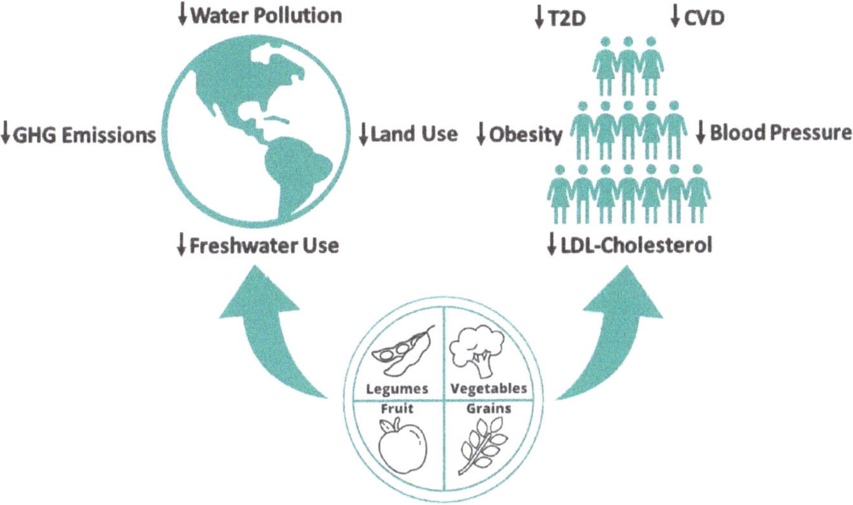

Figure 1. Summary of the planetary and human health benefits associated with the adoption of plant-based dietary patterns. Abbreviations: CVD, cardiovascular disease; GHG, greenhouse gas; LDL, low-density lipoprotein; T2D, type 2 diabetes.

2.2. Agricultural Land Use

Around 43% of the planet's ice-free terrestrial landmass is occupied by farmland (including croplands and pasturelands). Approximately 83% of this farmland is used to produce meat, eggs, farmed fish, and dairy, yet they only provide 18% and 37% of our calories and protein, respectively [9]. Per kilogram, animal products require more lifecycle energy inputs than plant foods [16]. The adoption of PBDs would substantially reduce agricultural land use. Eshel et al. [17] estimated that Americans could save approximately 34% and 24% of dietary and total land use, respectively, if they replaced all meat with plant-based alternatives. Considering the amount of land required to produce animal products, it is unsurprising that they are accountable for 67% of the deforestation caused by agriculture [9]. The destruction of ecosystems for croplands and pasturelands is the single largest factor causing species to be threatened with extinction [18]. Biodiversity is essential for the productivity and resilience of our food systems [19]. Shifting to PBDs would slow biodiversity loss substantially, thus having a protective effect on global food security [5]. It is estimated that animal product-free diets have the potential to reduce diet-related land use by 3.1 billion hectares (76% reduction), including a 19% reduction in arable land (Figure 1) [9].

2.3. Water Use

In total, 70% of all global freshwater withdrawals are used for the irrigation of crops, of which 24% are fed to livestock [5,20]. Approximately 43,000 L of water are required to produce 1 kg of beef, whereas it only takes 1000 L to produce 1 kg of grain [21]. A modelling study found that reducing animal product consumption would reduce global green and blue water use by 21% and 14%, respectively (Figure 1) [22]. PBDs may therefore play a role in water conservation. Animal product-free diets may also improve water quality by reducing eutrophication caused by nitrogenous fertilizer and manure runoff by 49% (Figure 1) [9].

2.4. Healthy Reference Dietary Pattern

The EAT-Lancet Commission has developed a healthy reference dietary pattern that would allow humanity to stay within a safe operating space, in terms of climate change, land use, biodiversity loss, freshwater use, and nitrogen and phosphorus pollution, even with a 10 billion global population [5]. The dietary pattern largely consists of fruits and vegetables, whole grains, legumes, nuts, and unsaturated oils; low to moderate consumption of seafood and poultry; zero to low consumption of red meat, processed meat, added sugar, refined grains, and starchy vegetables. Using data from the European Prospective Investigation into Cancer and Nutrition (EPIC) cohort involving 443,991 participants, Laine et al. [23] estimated that up to 19–63% of deaths and up to 10–39% of cancers could be prevented in a 20-year risk period by adopting different levels of adherence to the EAT-Lancet reference diet. They also estimated that switching from low adherence to higher adherence could reduce food-associated greenhouse gas emissions by up to 50% and land use up to 62%.

3. Human Health

Globally we are experiencing an unprecedented level of diet-related disease. Worldwide, 2.1 billion adults are overweight or obese [5]. Overweight and obesity are associated with a range of chronic diseases including type 2 diabetes (T2D) [24], hypertension [25], cardiovascular disease (CVD) [26], and some types of cancer [27]. Together, these diseases have a massive cost on society in terms of lives lost and healthcare spending. The Global Burden of Disease study estimated that increased consumption of whole grains, vegetables, nuts and seeds, and fruit could prevent 1.7 million, 1.8 million, 2.5 million, and 4.9 million premature deaths per year, respectively, via the beneficial effects on chronic disease risk factors [28].

3.1. Obesity

An extensive body of population studies and clinical trials supports the implementation of PBDs for the prevention of obesity and obesity-related diseases. Observational data from the Adventist Health Study-2 (AHS-2) involving 41,387 participants, showed that body mass index (BMI) was positively correlated with the amount of animal-based foods consumed, such that non-vegetarians had the highest BMI, followed by semi-vegetarians, pescatarians, vegetarians, and vegans [29]. In addition, findings from the EPIC-Oxford cohort, containing 21,966 men and women, have shown that vegans and pescatarian women gain significantly less weight annually compared with meat-eaters [30]. The lowest mean annual weight gain was observed in individuals who converted, during follow-up, to diets containing fewer animal-derived foods. In accordance with these findings, the European Prospective Investigation into Cancer, Physical Activity, Nutrition, Alcohol, Cessation of smoking, Eating out of home and obesity (EPIC-PANACEA) study found total meat consumption was positively associated with weight gain in 103,455 men and 270,348 women [31]. After adjusting for estimated energy intake, an additional 250 g/d of meat led to a 2 kg higher weight gain after 5 years (95% CI: 1.5, 2.7 kg). In a 5-year longitudinal study of 787 non-obese participants, dietary patterns were evaluated with overall plant-based diet index (PDI) scores, in which plant foods received positive scores and animal-derived foods received reverse scores [32]. A healthy PDI (hPDI) and an unhealthy PDI (uPDI) were also created. For the hPDI, healthy plant foods (fruits, vegetables, legumes, whole grains, nuts, and unsaturated vegetable oils) received positive scores, and animal foods and unhealthy plant foods (fruit juices, refined grains, and added sugars) received reverse scores. For the uPDI, unhealthy plant foods were allocated positive scores and animal foods and healthy plant foods were allocated reverse scores. At follow-up, both the hPDI (Risk Ratio (RR) = 0.31; 95% CI: 0.12–0.77) and overall PDI (RR = 0.56; 95% CI: 0.23–1.33) were inversely associated with obesity risk. However, only the hPDI association achieved statistical significance. Conversely, the uPDI was positively associated with obesity risk (RR = 1.94; 95% CI: 0.81–4.66); however, this finding was not statistically significant.

Robust evidence from clinical trials supports the use of PBDs for weight loss. In 2015, Barnard et al. [33] performed a meta-analysis of 15 clinical trials with vegan and vegetarian interventions lasting four weeks or more with no energy restrictions. Consumption of PBDs was associated with a mean weight change of −3.4 kg (95% CI: −4.4, −2.4 kg) in an intention-to-treat analysis and −4.6 kg (−5.4, −3.8 kg) in a completer analysis (Figure 1). Similarly, a 2021 meta-analysis of seven clinical trials found that PBDs significantly lowered bodyweight in Type 2 diabetics (−2.35 kg (95% CI: −3.51, −1.19)) [34]. A few new clinical trials assessing the effect of PBDs on bodyweight have been published since 2015 [35–40]. The BROAD study, which prescribed a whole food PBD, had noteworthy results [38]. It showed greater weight loss at 6 and 12 months than any other comparable interventional trial (no energy restrictions or regular exercise orders) to date.

3.2. Type 2 Diabetes

The global prevalence of T2D has nearly doubled in the past 30 years [41]. In 2021, diabetes was responsible for 6.7 million deaths and $966 billion USD in health expenditure [42] Large cohort studies show that the prevalence and incidence of T2D are significantly lower among those following PBDs. T2D prevalence in the AHS-2 cohort followed a similar trend as BMI with the lowest prevalence occurring in vegans (2.9%) and the highest in non-vegetarians (7.6%) [43]. Pescatarians (4.8%), semi-vegetarians (6.1%), and vegetarians (3.2%) had intermediate T2D prevalence. After adjusting for BMI and other confounding variables, vegans had half the risk of T2D than non-vegetarians (Odds Ratio (OR)) 0.51 (95% CI: 0.40, 0.66)) and semi-vegetarians had an intermediate risk (0.76 (0.65, 0.90)). In a 2-year prospective study of the AHS-2 cohort, vegans had less than half the risk of T2D than non-vegetarians (OR 0.38 (0.24, 0.62)) even when adjustments were made for BMI and other confounders [29]. In a 17-year prospective study with 8401 participants, long-term weekly dietary inclusion of meat was associated with 74% increased (OR 1.74 (1.36, 2.22)) odds of T2D compared with long-term adherence to a vegetarian dietary pattern [44]. Weekly meat intake remained an important risk factor (1.38 (1.06,1.68)) after adjusting for weight and weight change.

In a prospective study of three US cohorts (Nurses' Health Study (NHS), NHS II, Health Professionals Follow-up Study) totalling 192,657 participants, Chen et al. [45] evaluated the associations between changes in PBDs and subsequent T2D risk. During the 2,955,350 person-years of follow-up, 12,627 cases of T2D developed. Participant dietary patterns were evaluated with overall PDI, hPDI, and uPDI scores. Compared with participants whose indices remained stable over the 4-year follow-up, participants with the largest decrease (>10%) in PDI and hPDI had a 12–23% higher T2D risk in the subsequent 4 years. Each 10% increment in PDI and hPDI over 4 years was associated with a 7–9% lower T2D risk. It is worth noting that changes in the PDI scores were primarily due to changes in healthy plant-food intake, not changes in animal-derived food intake. No associations were observed between changes in uPDI and subsequent T2D risk. This may be due to the benefits of low animal food intake cancelling out the harmful effects associated with low intake of healthy plant foods [45].

A 2019 meta-analysis of nine prospective studies totalling 307,099 participants, found a significant inverse association between higher adherence to PBDs and T2D risk (RR 0.77 (95% CI: 0.71, 0.84)) in comparison with poorer adherence (Figure 1) [46]. As well as preventing T2D, there is evidence that PBDs may be an effective tool in the treatment of the disease. A meta-analysis of six controlled clinical trials found that consumption of PBDs was associated with a significant reduction in haemoglobin A1c (−0.39 points) compared with the consumption of omnivorous control diets [47]. This hypoglycaemic effect is approximately half of that observed with the prescription of the first-line medication, metformin [48].

3.3. CVD Risk

CVDs are the leading cause of mortality globally. In 2019, CVDs were responsible for 18.6 million deaths worldwide [49]. There is a range of evidence that supports the use of PBDs for the prevention of CVDs. A 2021 meta-analysis of prospective cohort studies totalling 698,707 participants, found that PBDs were associated with a 16% lower risk of CVD and an 11% lower risk of coronary heart disease (CHD) [50]. However, there were no associations between PBDs and risk of stroke. Another 2021 meta-analysis of prospective cohort studies totalling 410,085 participants found that PBDs were associated with a 10% lower risk of CVD incidence and 8% lower risk of cardiovascular mortality [51]. In a randomised secondary prevention trial (The Lyon Diet Heart Study) with 275 events recorded during a mean follow-up of 46 months, adherence to a plant-based Mediterranean-type dietary pattern was associated with a 72% reduction in cardiovascular events compared with adherence to a western-type dietary pattern [52]. In a randomised controlled trial with a 5-year follow-up, intensive lifestyle changes including the adoption of a healthful plant-based dietary pattern were shown to cause regression of atherosclerosis [53]. The control group in this trial had more than twice the risk of a cardiovascular event than the intensive lifestyle changes group (Figure 1). The reduced risk of CVD incidence and cardiovascular mortality observed in those following PBDs is likely due to the beneficial effects on CVD risk factors including overweight or obesity, T2D, hypertension, and hypercholesterolemia.

3.4. Hypertension and Hypercholesterolemia

In the AHS-2 cohort, vegans had approximately half the odds of hypertension than omnivores, even after controlling for BMI [54]. A 2021 meta-analysis including 41 controlled trials and 8416 participants found that PBDs significantly lower both systolic and diastolic blood pressure even with the inclusion of some animal products (Figure 1) [55]. A 2017 meta-analysis of 19 clinical trials including 1484 participants, found that compared with the consumption of omnivorous diets, vegetarian diets were significantly associated with decreased total cholesterol (-12.5 mg/dL) and low-density lipoprotein cholesterol (-12.2 mg/dL) (Figure 1) [56].

3.5. CVD Prevention

Taken together, the beneficial effects of PBDs on chronic disease risk factors found in controlled trials, and their associations with lower chronic disease risk found in prospective cohort studies provide strong support for the implementation of PBDs for chronic disease prevention. In a prospective cohort of 315,919 participants, high hPDI scores were associated with a 36% lower risk of mortality and each 10-point increase was associated with a 19% lower risk [57]. On the other hand, high uPDI scores were associated with a 41% increase in mortality risk and each 10-point increase was associated with a 15% increase in risk. This is supported by the most comprehensive systematic review on dietary patterns and all-cause mortality (ACM) to date [58]. It found that dietary patterns characterised by higher intake of vegetables, legumes, fruits, nuts, unrefined grains, fish, and unsaturated vegetable oils, and lower or no consumption of animal products (red and processed meat, meat and meat products, and high-fat dairy), refined grains, and added sugar, were associated with lower ACM risk.

4. Barriers and Potential Solutions

In 2020, a comprehensive review of the literature outlined the most prominent perceived and objective barriers preventing people from switching to PBDs [59]. The most prominent barrier to PBDs is meat appreciation and the difficulty perceived in abstaining from consumption (Figure 2). The development of plant-based meat alternatives provides an opportunity to overcome this barrier. Plant-based products have been developed to visually resemble meat and match the taste, structure, and nutritional value preferences of meat eaters. These products make the transition to PBDs less difficult and more appealing. Environmental life cycle assessments for two popular plant-based substitutes, Beyond

Meat's Beyond Burger and Impossible Food's Impossible Burger, showed that switching from beef to either of the products reduces GHG emissions, land use, and water footprint by approximately 90% [60,61]. Although plant-based meat alternatives are classified as ultra-processed, they may still exert some of the beneficial effects on CVD risk factors as healthy PBDs [62]. A randomized cross-over trial investigating the effect of Beyond Meat products versus animal-derived meat on CVD risk factors found that consumption of plant-based meat alternatives was associated with significantly lower trimethylamine-N-oxide (TMAO) concentrations, LDL-cholesterol concentrations, and body weight compared with the consumption of animal meat [63]. Moreover, there were no adverse effects on other risk factors during the plant-based phase. More controlled trials are needed to characterize the effect of ultra-processed meat analogues on health markers.

Figure 2. The main barriers to widespread adoption of plant-based dietary patterns.

The second most prominent barrier to PBDs is health concerns, specifically nutrient deficiencies, for example, protein and calcium (Figure 2) [59]. International and national commitments to PBDs demonstrated by investment in public health and sustainability education could break down these barriers. The public needs to be educated on specific plant-based food sources of essential nutrients such as iron, calcium, and zinc and be reassured that their protein needs can be sufficiently met. A potential strategy for relieving the perceived health concerns attached to PBDs is to provide proper nutrition education to medical students and health professionals. A survey of medical schools found that on average fewer than 20 h over four years are spent on nutrition education [64]. Accordingly, physicians often lack important nutrition knowledge and the counselling skills required to successfully guide their patients [65–75]. In a survey of resident physicians, only 14% of participants felt physicians were adequately trained to provide nutritional counselling [76]. Ironically, in a survey of the public, 61% of participants considered physicians to be "very credible" sources of nutrition information [77]. Educating doctors on how to prevent and treat chronic diseases with healthful PBDs may have positive effects beyond individual patient care, by influencing the wider public's negative perceptions of PBDs. However, a lack of nutrition training is not the only way that physicians act as barriers. Firstly, they may have conflicts of interest and personal prejudices that bias their views on PBDs, preventing them from promoting the implementation of PBDs. Secondly, there is a lack of financial incentive for physicians to implement the use of PBDs [78,79]. Preventing chronic diseases with healthful PBDs reduces the demand for expensive medical treatments and procedures, which results in reduced income for physicians.

The third most common barrier relates to convenience and tastes factors (Figure 2) [59]. The availability of plant-based options out of home are limited and people believe that the preparation of plant-based meals is complicated. PBDs are also perceived as tasteless [80]. New policies mandating that canteens at schools, hospitals, universities, and other state-owned services must provide healthful plant-based options could be implemented to reduce the convenience barrier. Incentives for businesses to offer more healthful plant-based options would also help to overcome this barrier. Online educational resources and community cooking classes could be utilized to facilitate the teaching of plant-based food preparation to the public, potentially tacking both convenience and taste factors [79]. Taste barriers could also be overcome with the previously mentioned meat analogues.

The final prominent barrier to PBDs is the expense of plant-based foods (Figure 2) [59]. This barrier could be broken down by allocating subsidies to the production of sustainable, healthful foods (e.g., fruits and vegetables) financed by a tax on unhealthful, environmentally damaging foods (e.g., red and processed meat) or an incremental increase in income tax [81]. It is estimated that a subsidy of 25% of the cost of fruits and vegetables could close the gap between the recommended intake and the actual average intake by a third [81].

Author Contributions: J.G. and F.P.C. conceptualized the article, J.G. wrote the manuscript, F.P.C. contributed to the revision. All authors have read and agreed to the published version of the manuscript.

Funding: J.G. is supported by an Economic & Social Research Council PhD Scholarship. This research received no external funding.

Institutional Review Board Statement: Not applicable.

Informed Consent Statement: Not applicable.

Data Availability Statement: Not applicable.

Acknowledgments: The views expressed herein are not necessarily the views or the stated policy of the World Health Organization (WHO) and the presentation of the material does not imply the expression of any opinion on the part of WHO.

Conflicts of Interest: F.P.C.: Past president, the British and Irish Hypertension Society (2017-9) (unpaid); Member, Action on Salt and World Action on Salt, Sugar and Health (unpaid); Head, World Health Organization (WHO) Collaborating Centre for Nutrition (unpaid); Senior Advisor, WHO (received travel, accommodation, per-diem, refund of expenses); OMRON Academy (received speaker fees, travel, accommodation, expenses); annual royalties from Oxford University Press (OUP) for two books on topics unrelated to salt. J.G. declares no conflict of interest.

References

1. Stocker, T.F.; Qin, D.; Plattner, G.-K.; Tignor, M.; Allen, S.K.; Boschung, J.; Nauels, A.; Xia, Y.; Bex, B.; Midgley, B.M. *IPCC, 2013: Climate Change 2013: The Physical Science Basis. Contribution of Working Group I to the Fifth Assessment Report of the Intergovernmental Panel on Climate Change*; Cambridge University Press: Cambridge, UK; New York, NY, USA, 2013.
2. Horowitz, C.A. Paris agreement. *Int. Leg. Mater.* **2016**, *55*, 740–755. [CrossRef]
3. Schwab, K.; Malleret, T. *The Great Reset*; World Economic Forum: Geneva, Switzerland, 2020.
4. Johnson, B. *The Ten Point Plan for a Green Industrial Revolution*; H.M. Government: London, UK, 2020.
5. Willett, W.; Rockström, J.; Loken, B.; Springmann, M.; Lang, T.; Vermeulen, S.; Garnett, T.; Tilman, D.; DeClerck, F.; Wood, A. Food in the Anthropocene: The EAT–Lancet Commission on healthy diets from sustainable food systems. *Lancet* **2019**, *393*, 447–492. [CrossRef]
6. Vermeulen, S.J.; Campbell, B.M.; Ingram, J.S.I. Climate change and food systems. *Annu. Rev. Environ. Resour.* **2012**, *37*, 195–222. [CrossRef]
7. Steffen, W.; Richardson, K.; Rockström, J.; Cornell, S.E.; Fetzer, I.; Bennett, E.M.; Biggs, R.; Carpenter, S.R.; De Vries, W.; De Wit, C.A. Planetary boundaries: Guiding human development on a changing planet. *Science* **2015**, *347*, 1259855. [CrossRef]
8. Foley, J.A.; DeFries, R.; Asner, G.P.; Barford, C.; Bonan, G.; Carpenter, S.R.; Chapin, F.S.; Coe, M.T.; Daily, G.C.; Gibbs, H.K. Global consequences of land use. *Science* **2005**, *309*, 570–574. [CrossRef]
9. Poore, J.; Nemecek, T. Reducing food's environmental impacts through producers and consumers. *Science* **2018**, *360*, 987–992. [CrossRef]
10. Hemler, E.C.; Hu, F.B. Plant-based diets for personal, population, and planetary health. *Adv. Nutr.* **2019**, *10*, S275–S283. [CrossRef]
11. Fresán, U.; Sabaté, J. Vegetarian diets: Planetary health and its alignment with human health. *Adv. Nutr.* **2019**, *10*, S380–S388. [CrossRef]
12. Wilson, N.; Cleghorn, C.L.; Cobiac, L.J.; Mizdrak, A.; Nghiem, N. Achieving healthy and sustainable diets: A review of the results of recent mathematical optimization studies. *Adv. Nutr.* **2019**, *10*, S389–S403. [CrossRef]

13. Shukla, P.R.; Skeg, J.; Buendia, E.C.; Masson-Delmotte, V.; Pörtner, H.O.; Roberts, D.C.; Zhai, P.; Slade, R.; Connors, S.; van Diemen, S. *Climate Change and Land: An IPCC Special Report on Climate Change, Desertification, Land Degradation, Sustainable Land Management, Food Security, and Greenhouse Gas Fluxes in Terrestrial Ecosystems*; IPCC: Geneva, Switzerland, 2019.

14. Romanello, M.; McGushin, A.; Di Napoli, C.; Drummond, P.; Hughes, N.; Jamart, L.; Kennard, H.; Lampard, P.; Rodriguez, B.S.; Arnell, N. The 2021 report of the Lancet Countdown on health and climate change: Code red for a healthy future. *Lancet* **2021**, *398*, 1619–1662. [CrossRef]

15. Lynch, H.; Johnston, C.; Wharton, C. Plant-based diets: Considerations for environmental impact, protein quality, and exercise performance. *Nutrients* **2018**, *10*, 1841. [CrossRef]

16. Carlsson-Kanyama, A.; Ekström, M.P.; Shanahan, H. Food and life cycle energy inputs: Consequences of diet and ways to increase efficiency. *Ecol. Econ.* **2003**, *44*, 293–307. [CrossRef]

17. Eshel, G.; Stainier, P.; Shepon, A.; Swaminathan, A. Environmentally optimal, nutritionally sound, protein and energy conserving plant based alternatives to US meat. *Sci. Rep.* **2019**, *9*, 10345. [CrossRef]

18. Tilman, D.; Clark, M.; Williams, D.R.; Kimmel, K.; Polasky, S.; Packer, C. Future threats to biodiversity and pathways to their prevention. *Nature* **2017**, *546*, 73–81. [CrossRef] [PubMed]

19. Cardinale, B.J.; Duffy, J.E.; Gonzalez, A.; Hooper, D.U., Perringe, C.; Venail, P.; Narwani, A.; Mace, G.M.; Tilman, D.; Wardle, D.A. Biodiversity loss and its impact on humanity. *Nature* **2012**, *486*, 59–67. [CrossRef]

20. Cassidy, E.S.; West, P.C.; Gerber, J.S.; Foley, J.A. Redefining agricultural yields: From tonnes to people nourished per hectare. *Environ. Res. Lett.* **2013**, *8*, 034015. [CrossRef]

21. Pimentel, D.; Berger, B.; Filiberto, D.; Newton, M.; Wolfe, B.; Karabinakis, E.; Clark, S.; Poon, E.; Abbett, E.; Nandagopal, S. Water resources: Agricultural and environmental issues. *BioScience* **2004**, *54*, 909–918. [CrossRef]

22. Jalava, M.; Kummu, M.; Porkka, M.; Siebert, S.; Varis, O. Diet change—A solution to reduce water use? *Environ. Res. Lett.* **2014**, *9*, 074016. [CrossRef]

23. Laine, J.E.; Huybrechts, I.; Gunter, M.J.; Ferrari, P.; Weiderpass, E.; Tsilidis, K.; Aune, D.; Schulze, M.B.; Bergmann, M.; Temme, E.H.M. Co-benefits from sustainable dietary shifts for population and environmental health: An assessment from a large European cohort study. *Lancet Planet. Health* **2021**, *5*, e786–e796. [CrossRef]

24. Smyth, S.; Heron, A. Diabetes and obesity: The twin epidemics. *Nat. Med.* **2006**, *12*, 75–80. [CrossRef] [PubMed]

25. Rahmouni, K.; Correia, M.L.G.; Haynes, W.G.; Mark, A.L. Obesity-associated hypertension: New insights into mechanisms. *Hypertension* **2005**, *45*, 9–14. [CrossRef]

26. Van Gaal, L.F.; Mertens, I.L.; De Block, C.E. Mechanisms linking obesity with cardiovascular disease. *Nature* **2006**, *444*, 875–880. [CrossRef]

27. Wolin, K.Y.; Carson, K.; Colditz, G.A. Obesity and cancer. *Oncologist* **2010**, *15*, 556–565. [CrossRef]

28. Lim, S.S.; Vos, T.; Flaxman, A.D.; Danaei, G.; Shibuya, K.; Adair-Rohani, H.; Almazroa, M.A.; Amann, M.; Anderson, H.R.; Andrews, K.G.; et al. A comparative risk assessment of burden of disease and injury attributable to 67 risk factors and risk factor clusters in 21 regions, 1990–2010: A systematic analysis for the Global Burden of Disease Study 2010. *Lancet* **2012**, *380*, 2224–2260. [CrossRef]

29. Tonstad, S.; Stewart, K.; Oda, K.; Batech, M.; Herring, R.P.; Fraser, G.E. Vegetarian diets and incidence of diabetes in the Adventist Health Study-2. *Nutr. Metab. Cardiovasc. Dis.* **2013**, *23*, 292–299. [CrossRef]

30. Rosell, M.; Appleby, P.; Spencer, E.; Key, T. Weight gain over 5 years in 21,966 meat-eating, fish-eating, vegetarian, and vegan men and women in EPIC-Oxford. *Int. J. Obes.* **2006**, *30*, 1389–1396. [CrossRef]

31. Vergnaud, A.-C.; Norat, T.; Romaguera, D.; Mouw, T.; May, A.M.; Travier, N.; Luan, J.a.; Wareham, N.; Slimani, N.; Rinaldi, S. Meat consumption and prospective weight change in participants of the EPIC-PANACEA study. *Am. J. Clin. Nutr.* **2010**, *92*, 398–407. [CrossRef]

32. Wang, Y.B.; Shivappa, N.; Hébert, J.R.; Page, A.J.; Gill, T.K.; Melaku, Y.A. Association between dietary inflammatory index, dietary patterns, plant-based dietary index and the risk of obesity. *Nutrients* **2021**, *13*, 1536. [CrossRef]

33. Barnard, N.D.; Levin, S.M.; Yokoyama, Y. A systematic review and meta-analysis of changes in body weight in clinical trials of vegetarian diets. *J. Acad. Nutr. Dict.* **2015**, *115*, 954–969. [CrossRef]

34. Austin, G.; Ferguson, J.J.A.; Garg, M.L. Effects of Plant-Based Diets on Weight Status in Type 2 Diabetes: A Systematic Review and Meta-Analysis of Randomised Controlled Trials. *Nutrients* **2021**, *13*, 4099. [CrossRef]

35. Turner-McGrievy, G.M.; Davidson, C.R.; Wingard, E.E.; Wilcox, S.; Frongillo, E.A. Comparative effectiveness of plant-based diets for weight loss: A randomized controlled trial of five different diets. *Nutrition* **2015**, *31*, 350–358. [CrossRef] [PubMed]

36. Turner-McGrievy, G.M.; Davidson, C.R.; Wingard, E.E.; Billings, D.L. Low glycemic index vegan or low-calorie weight loss diets for women with polycystic ovary syndrome: A randomized controlled feasibility study. *Nutr. Res.* **2014**, *34*, 552–558. [CrossRef] [PubMed]

37. Talreja, A.; Talreja, S.; Talreja, R.; Talreja, D. CRT-601 The VA beach diet study: An investigation of the effects of plant-based, mediterranean, paleolithic, and dash diets on cardiovascular disease risk. *JACC Cardiovasc. Interv.* **2015**, *8*, S41. [CrossRef]

38. Wright, N.; Wilson, L.; Smith, M.; Duncan, B.; McHugh, P. The BROAD study: A randomised controlled trial using a whole food plant-based diet in the community for obesity, ischaemic heart disease or diabetes. *Nutr. Diabetes* **2017**, *7*, e256. [CrossRef]

39. Kahleova, H.; Fleeman, R.; Hlozkova, A.; Holubkov, R.; Barnard, N.D. A plant-based diet in overweight individuals in a 16-week randomized clinical trial: Metabolic benefits of plant protein. *Nutr. Diabetes* **2018**, *8*, 58. [CrossRef]

40. Barnard, N.D.; Levin, S.M.; Gloede, L.; Flores, R. Turning the waiting room into a classroom: Weekly classes using a vegan or a portion-controlled eating plan improve diabetes control in a randomized translational study. *J. Acad. Nutr. Diet.* **2018**, *118*, 1072–1079. [CrossRef]

41. Zhou, B.; Lu, Y.; Hajifathalian, K.; Bentham, J.; Di Cesare, M.; Danaei, G.; Bixby, H.; Cowan, M.J.; Ali, M.K.; Taddei, C. Worldwide trends in diabetes since 1980: A pooled analysis of 751 population-based studies with 4·4 million participants. *Lancet* **2016**, *387*, 1513–1530. [CrossRef]

42. Sun, H.; Saeedi, P.; Karuranga, S.; Pinkepank, M.; Ogurtsova, K.; Duncan, B.B.; Stein, C.; Basit, A.; Chan, J.C.N.; Mbanya, J.C. IDF diabetes atlas: Global, regional and country-level diabetes prevalence estimates for 2021 and projections for 2045. *Diabetes Res. Clin. Pract.* **2022**, *183*, 109119. [CrossRef]

43. Tonstad, S.; Butler, T.; Yan, R.; Fraser, G.E. Type of Vegetarian Diet, Body Weight, and Prevalence of Type 2 Diabetes. *Diabetes Care* **2009**, *32*, 791–796. [CrossRef] [PubMed]

44. Vang, A.; Singh, P.N.; Lee, J.W.; Haddad, E.H.; Brinegar, C.H. Meats, processed meats, obesity, weight gain and occurrence of diabetes among adults: Findings from Adventist Health Studies. *Ann. Nutr. Metab.* **2008**, *52*, 96–104. [CrossRef]

45. Chen, Z.; Drouin-Chartier, J.-P.; Li, Y.; Baden, M.Y.; Manson, J.E.; Willett, W.C.; Voortman, T.; Hu, F.B.; Bhupathiraju, S.N. Changes in plant-based diet indices and subsequent risk of type 2 diabetes in women and men: Three US prospective cohorts. *Diabetes Care* **2021**, *44*, 663–671. [CrossRef] [PubMed]

46. Qian, F.; Liu, G.; Hu, F.B.; Bhupathiraju, S.N.; Sun, Q. Association between plant-based dietary patterns and risk of type 2 diabetes: A systematic review and meta-analysis. *JAMA Intern. Med.* **2019**, *179*, 1335–1344. [CrossRef] [PubMed]

47. Yokoyama, Y.; Barnard, N.D.; Levin, S.M.; Watanabe, M. Vegetarian diets and glycemic control in diabetes: A systematic review and meta-analysis. *Cardiovasc. Diagn. Ther.* **2014**, *4*, 373. [PubMed]

48. Johansen, K. Efficacy of metformin in the treatment of NIDDM. Meta-analysis. *Diabetes Care* **1999**, *22*, 33–37. [CrossRef] [PubMed]

49. Roth, G.A.; Mensah, G.A.; Johnson, C.O.; Addolorato, G.; Ammirati, E.; Baddour, L.M.; Barengo, N.C.; Beaton, A.Z.; Benjamin, E.J.; Benziger, C.P. Global burden of cardiovascular diseases and risk factors, 1990–2019: Update from the GBD 2019 study. *J. Am. Coll. Cardiol.* **2020**, *76*, 2982–3021. [CrossRef]

50. Gan, Z.H.; Cheong, H.C.; Tu, Y.-K.; Kuo, P.-H. Association between Plant-Based Dietary Patterns and Risk of Cardiovascular Disease: A Systematic Review and Meta-Analysis of Prospective Cohort Studies. *Nutrients* **2021**, *13*, 3952. [CrossRef]

51. Quek, J.; Lim, G.; Lim, W.H.; Ng, C.H.; So, W.Z.; Toh, J.; Pan, X.H.; Chin, Y.H.; Muthiah, M.D.; Chan, S.P. The Association of Plant-Based Diet With Cardiovascular Disease and Mortality: A Meta-Analysis and Systematic Review of Prospect Cohort Studies. *Front. Cardiovasc. Med.* **2021**, *8*, 756810. [CrossRef]

52. De Lorgeril, M.; Salen, P.; Martin, J.-L.; Monjaud, I.; Delaye, J.; Mamelle, N. Mediterranean diet, traditional risk factors, and the rate of cardiovascular complications after myocardial infarction: Final report of the Lyon Diet Heart Study. *Circulation* **1999**, *99*, 779–785. [CrossRef]

53. Ornish, D.; Scherwitz, L.W.; Billings, J.H.; Gould, K.L.; Merritt, T.A.; Sparler, S.; Armstrong, W.T.; Ports, T.A.; Kirkeeide, R.L.; Hogeboom, C. Intensive lifestyle changes for reversal of coronary heart disease. *JAMA* **1998**, *280*, 2001–2007. [CrossRef]

54. Pettersen, B.J.; Anousheh, R.; Fan, J.; Jaceldo-Siegl, K.; Fraser, G.E. Vegetarian diets and blood pressure among white subjects: Results from the Adventist Health Study-2 (AHS-2). *Public Health Nutr.* **2012**, *15*, 1909–1916. [CrossRef] [PubMed]

55. Gibbs, J.; Gaskin, E.; Ji, C.; Miller, M.A.; Cappuccio, F.P. The effect of plant-based dietary patterns on blood pressure: A systematic review and meta-analysis of controlled intervention trials. *J. Hypertens.* **2021**, *39*, 23–37. [CrossRef]

56. Yokoyama, Y.; Levin, S.M.; Barnard, N.D. Association between plant-based diets and plasma lipids: A systematic review and meta-analysis. *Nutr. Rev.* **2017**, *75*, 683–698. [CrossRef]

57. Wang, D.D.; Li, Y.; Nguyen, X.-M.T.; Song, R.J.; Ho, Y.-L.; Hu, F.B.; Willett, W.C.; Wilson, P.W.F.; Cho, K.; Gaziano, J.M. Degree of Adherence to Based Diet and Total and Cause-Specific Mortality: Prospective Cohort Study in the Million Veteran Program. *Public Health Nutr.* **2022**, *25*, 1–38. [CrossRef]

58. English, L.K.; Ard, J.D.; Bailey, R.L.; Bates, M.; Bazzano, L.A.; Boushey, C.J.; Brown, C.; Butera, G.; Callahan, E.H.; De Jesus, J. Evaluation of dietary patterns and all-cause mortality: A systematic review. *JAMA Netw. Open* **2021**, *4*, e2122277. [CrossRef]

59. Fehér, A.; Gazdecki, M.; Véha, M.; Szakály, M.; Szakály, Z. A Comprehensive Review of the Benefits of and the Barriers to the Switch to a Plant-Based Diet. *Sustainability* **2020**, *12*, 4136. [CrossRef]

60. Heller, M.C.; Keoleian, G.A. *Beyond Meat's Beyond Burger Life Cycle Assessment: A Detailed Comparison between a Plant-Based and an Animal-Based Protein Source*; University of Michigan: Michigan, MI, USA, 2018.

61. Khan, S.; Dettling, J.; Hester, J.; Moses, R.; Foods, I. *Comparative Environmental LCA of the Impossible Burger with Conventional Ground Beef Burger*; Quantis: Lausanne, Switzerland, 2019.

62. Bohrer, B.M. An investigation of the formulation and nutritional composition of modern meat analogue products. *Food Sci. Hum. Wellness* **2019**, *8*, 320–329. [CrossRef]

63. Crimarco, A.; Springfield, S.; Petlura, C.; Streaty, T.; Cunanan, K.; Lee, J.; Fielding-Singh, P.; Carter, M.M.; Topf, M.A.; Wastyk, H.C. A randomized crossover trial on the effect of plant-based compared with animal-based meat on trimethylamine-N-oxide and cardiovascular disease risk factors in generally healthy adults: Study With Appetizing Plantfood—Meat Eating Alternative Trial (SWAP-MEAT). *Am. J. Clin. Nutr.* **2020**, *112*, 1188–1199.

64. Adams, K.M.; Kohlmeier, M.; Zeisel, S.H. Nutrition education in US medical schools: Latest update of a national survey. *Acad. Med. J. Assoc. Am. Med. Coll.* **2010**, *85*, 1537. [CrossRef]

65. Chung, M.; Van Buul, V.J.; Wilms, E.; Nellessen, N.; Brouns, F. Nutrition education in European medical schools: Results of an international survey. *Eur. J. Clin. Nutr.* **2014**, *68*, 844–846. [CrossRef]
66. Dumic, A.; Miskulin, M.; Pavlovic, N.; Orkic, Z.; Bilic-Kirin, V.; Miskulin, I. The nutrition knowledge of Croatian general practitioners. *J. Clin. Med.* **2018**, *7*, 178. [CrossRef]
67. Grammatikopoulou, M.G.; Katsouda, A.; Lekka, K.; Tsantekidis, K.; Bouras, E.; Kasapidou, E.; Poulia, K.-A.; Chourdakis, M. Is continuing medical education sufficient? Assessing the clinical nutrition knowledge of medical doctors. *Nutrition* **2019**, *57*, 69–73. [CrossRef] [PubMed]
68. Hyska, J.; Mersini, E.; Mone, I.; Bushi, E.; Sadiku, E.; Hoti, K.; Bregu, A. Assessment of knowledge, attitudes and practices about public health nutrition among students of the University of Medicine in Tirana, Albania. *S. East. Eur. J. Public Health (SEEJPH)* **2015**, I. [CrossRef]
69. Devries, S.; Agatston, A.; Aggarwal, M.; Aspry, K.E.; Esselstyn, C.B.; Kris-Etherton, P.; Miller, M.; O'Keefe, J.H.; Ros, E.; Rzeszut, A.K. A deficiency of nutrition education and practice in cardiology. *Am. J. Med.* **2017**, *130*, 1298–1305. [CrossRef] [PubMed]
70. Devries, S.; Dalen, J.E.; Eisenberg, D.M.; Maizes, V.; Ornish, D.; Prasad, A.; Sierpina, V.; Weil, A.T.; Willett, W. A deficiency of nutrition education in medical training. *Am. J. Med.* **2014**, *127*, 804–806. [CrossRef]
71. Crowley, J.; Ball, L.; Hiddink, G.J. Nutrition in medical education: A systematic review. *Lancet Planet. Health* **2019**, *3*, e379–e389. [CrossRef]
72. Sanne, I.; Bjørke-Monsen, A.-L. Lack of nutritional knowledge among Norwegian medical students concerning vegetarian diets. *J. Public Health* **2020**, *30*, 495–501. [CrossRef]
73. Bettinelli, M.E.; Bezze, E.; Morasca, L.; Plevani, L.; Sorrentino, G.; Morniroli, D.; Giannì, M.L.; Mosca, F. Knowledge of health professionals regarding vegetarian diets from pregnancy to adolescence: An observational study. *Nutrients* **2019**, *11*, 1149. [CrossRef]
74. Hamiel, U.; Landau, N.; Fuhrer, A.E.; Shalem, T.; Goldman, M. The knowledge and attitudes of pediatricians in Israel towards vegetarianism. *J. Pediatric Gastroenterol. Nutr.* **2020**, *71*, 119–124. [CrossRef]
75. Nestle, M.; Baron, R.B. Nutrition in medical education: From counting hours to measuring competence. *JAMA Intern. Med.* **2014**, *174*, 843–844. [CrossRef]
76. Vetter, M.L.; Herring, S.J.; Sood, M.; Shah, N.R.; Kalet, A.L. What do resident physicians know about nutrition? An evaluation of attitudes, self-perceived proficiency and knowledge. *J. Am. Coll. Nutr.* **2008**, *27*, 287–298. [CrossRef]
77. American Dietetic Association. *Nutrition and You: Trends 2008*; ADA: Chicago, IL, USA, 2008.
78. Storz, M. Barriers to the Plant-Based Movement: A Physician's Perspective. *Int. J. Dis. Reversal Prev.* **2020**, *2*, 27–30. [CrossRef]
79. Jakše, B. Placing a Well-Designed Vegan Diet for Slovenes. *Nutrients* **2021**, *13*, 4545. [CrossRef] [PubMed]
80. Lea, E.; Worsley, A. Influences on meat consumption in Australia. *Appetite* **2001**, *36*, 127–136. [CrossRef] [PubMed]
81. Pancrazi, R.; van Rens, T.; Vukotić, M. How distorted food prices discourage a healthy diet. *Sci. Adv.* **2022**, *8*, eabi8807. [CrossRef] [PubMed]

nutrients

MDPI

Article

School Fruit and Vegetables Scheme: Characteristics of Its Implementation in the European Union from 2009/10 to 2016/17

Iris Comino [1,2], Panmela Soares [1,2], María Asunción Martínez-Milán [1], Pablo Caballero [1,*] and María Carmen Davó-Blanes [1,2]

1   Department of Community Nursing, Preventive Medicine and Public Health and History of Science, University of Alicante, 03690 Alicante, Spain; iriscomino@gmail.com (I.C.); panmela.soares@ua.es (P.S.); mariasuncion.m.m@gmail.com (M.A.M.-M.); mdavo@ua.es (M.C.D.-B.)
2   Public Health Research Group, Department of Community Nursing, Preventive Medicine and Public Health and History of Science, University of Alicante, 03690 Alicante, Spain
*   Correspondence: pablo.caballero@ua.es

Abstract: The "School Fruit and Vegetables Scheme" (SFVS) was proposed in 2009/10 as a strategy to support the consumption of Fruit and Vegetables (FV), decrease rates of obesity, improve agricultural income, stabilize markets, and ensure the current and future supply of these foods. However, there is little information about how it was carried out in the EU. Given the potential of the SFVS to support healthier, more sustainable food systems, the objective of this study was to identify the characteristics of SFVS implementation from 2009/10 to 2016/17 in the EU. A longitudinal, observational, and retrospective study was carried out based on secondary data. A total of 186 annual reports of the Member States (MS) participating in the SFVS from 2009/10 to 2016/17 were consulted: European and national budget, funds used from the EU, participating schools and students, duration of the SFVS, FV offered, and application of sustainability criteria, expenditure per student, days of the week, the quantity of FV offered per student and other indicators were calculated. The majority of MS participated in the SFVS during the study period with a heterogeneous implementation pattern in terms of funds used, coverage, duration, quantity (totals and by portion), and cost of FV distributed per student. The sustainability criteria for the FV distribution were also not applied uniformly in all the MS. Establishing minimum recommendations for SFVS implementation are recommended to maximize the benefits of the SFVS. The results may be useful for planning new strategies to help address and improve current health and environmental problems.

Keywords: schoolchildren; fruit and vegetables; community nutrition; nutrition interventions; health promotion

Citation: Comino, I.; Soares, P.; Martínez-Milán, M.A.; Caballero, P.; Davó-Blanes, M.C. School Fruit and Vegetables Scheme: Characteristics of Its Implementation in the European Union from 2009/10 to 2016/17. *Nutrients* **2022**, *14*, 3069. https://doi.org/10.3390/nu14153069

Academic Editors: Colin Bell and Penny Love

Received: 21 June 2022
Accepted: 22 July 2022
Published: 26 July 2022

Publisher's Note: MDPI stays neutral with regard to jurisdictional claims in published maps and institutional affiliations.

1. Introduction

The low consumption of fruit and vegetables (FV) is an avoidable risk factor that contributes to an increase in non-transmissible diseases [1,2]. In 2017, 3.9 million deaths around the world were attributed to inadequate intake of FV [3]. Low consumption of FV in the European Union (EU) continues to be a problem related to social, physical, and economic accessibility [4]. Despite the efforts of governments to promote FV consumption, surveys of the child population indicate that average consumption is below recommended levels [5–9].

Considering that dietary habits established in childhood and adolescence tend to extend through adult ages [10,11], supporting the consumption of FV in children is a public health priority [6]. Diverse strategies have been carried out in school environments with this aim [12–14]. Available evidence suggests that school feeding programs are effective in improving the consumption of FV [15–17], specifically, those that promote the availability and accessibility of FV via distribution in schools [18–24].

Under the Common Agricultural Policy (CAP), the European Commission (EC) proposed increasing financial funding to support the consumption of FV in the school population [25]. The "School Fruit and Vegetables Scheme" (SFVS) was proposed in 2009/10 as a strategy to support the consumption of FV, decrease rates of obesity, improve agricultural income, stabilize markets, and ensure the current and future supply of these foods [26]. The SFVS consisted of European Union (EU) subsidized distribution of FV in pre-school centers, primary and secondary schools, and outside of school canteen hours. Furthermore, the SFVS included educational measures (EM), whose objective was to raise awareness among school children, connecting them not only to their health and well-being but also to agriculture, food production, and environmental sustainability issues. Along the lines of the Sustainable Development Goals (SDG), the SFVS recommended that distributed foods be organic, seasonal, and from local producers originating in the European Community. This is especially important considering the effect of the production of these foods on the food system [27].

From its application in 2009/10 until 2016/17 (the year before the implementation of the EU School Fruit, Vegetables and Milk Scheme), the SFVS benefitted from a budget of approximately 937,226,513.40 Euros. Member States (MS) interested in participating in the SFVS must develop an annual pre-strategy at the national or regional level. The overall target group of the scheme (from kindergartens to secondary schools) is set by the EU and MS choose their target group from this age range. In addition, MS must provide national funds to accompany EU support for the purchase and distribution of FV, the implementation of EM, and the monitoring and evaluation of the SFVS [26]. Although the SFVS was implemented in 2009/10, there is little information about how it was carried out in the EU. Given the potential of the SFVS to support healthier, more sustainable food systems, the objective of this study was to identify the characteristics of SFVS implementation from 2009/10 to 2016/17 in the EU.

2. Materials and Methods

A longitudinal, observational, retrospective study based on public secondary data was conducted and therefore no ethical approval was required. During the study period, between 21 and 25 of the 28 EU Member States (before Brexit in 2020) participated in the SFVS: Austria (AUT), Belgium (BEL), Bulgaria (BRG), Croatia (HRV), Cyprus (CYP), Czech Republic (CZE), Denmark (DNK), Estonia (EST), France (FRA), Germany (DEU), Greece (GRC), Hungary (HUN), Ireland (IRL), Italy (ITA), Latvia (LVA), Lithuania (LTU), Luxembourg (LUX), Malta (MLT), Poland (POL), Portugal (PRT), Romania (ROU), Slovakia (SVK), Slovenia (SVN), Spain (ESP), The Netherlands (NLD). Between 2017 and 2018, the annual reports of the 25 MS participating in the SFVS from 2009/10 to 2016/17 were obtained from the EC website, i.e., 186 reports in total [26].

With the information contained in the reports, by school year and MS, the database was constructed including the following variables: European and national budget (€), funds used from the EU (%), participating schools (n, %), participating students (n, %), duration of the SFVS (weeks, days of the week, days of FV deliveries), the quantity of FV offered (t, portions per participant), and application of sustainability criteria (local, seasonal, organic and community origin (%)). In addition, the following variables were calculated: expenditure per student (€), days of the week on which the SFVS was implemented, the quantity of FV offered per student (in kg), average portions offered per student, and the average price per portion. To calculate average participation (both for schools and for students), values were calculated for each MS during the middle of the period for the 2013 school year. To calculate the percentage of countries that applied the recommendation on local foods, seasonal foods, organic foods, and community-origin foods, the MS that provided data in their reports were used. To calculate the expenditure per student, the budget for each MS was divided between the students that complied with the SFVS in each country. Some data for these variables were missing but could be calculated from other data available in the reports. In no case have estimation methods been used for missing data. All indicators have been calculated with the available data (see Supplementary Material).

For the analysis, the described variables were organized into four sections: two for the total data of the EU, and another two with the data from the different MS, organized into eight sub-sections.

Sections for the study of SFVS implementation for the European Union globally:

General Characteristics of the SFVS by School Year for the EU As a Whole: Includes MS participants, budget (%), expenditure per student (€), % participation (schools and students), and duration (weeks and days of FV deliveries).

Quantity of FV Included in the SFVS by School Year and for the EU As a Whole: Includes the quantity of FV purchased/distributed (tons), FV per student (kg), average portions per student (n), weight per portion (g) and price per portion (€).

Sections and sub-sections for the study of the implementation of the SFVS for participating MS:

Involvement of the MS in Carrying out the SFVS: Included four sub-sections:

Economic investment in the SFVS by MS during the study period: Included EU funds used by each MS (%) and expenditure per student (€).

SFVS coverage by MS during the study period: Percentage of students covered by FV (%) and percentage of students covered by EM (%).

Days of FV deliveries and quantity of FV per student (kg) by MS during the study period.

Economic investment in terms of the coverage, duration, and quantity of FV distributed by the SFVS.

Characteristics of the Distribution of FV Among Students: Included four sub-sections:

Frequency of FV deliveries: Duration of the SFVS in weeks and days of FV deliveries

Quantities of FV distributed and percentage of students covered by duration. Relationships between FV per student (kg), duration (days of FV deliveries), and percentage of students covered (%).

Portions distributed per student. Relationships between price (€), weight (g), and the number of portions (n).

Sustainability criteria applied in the distribution of FV: included local, seasonal, organic and community origin criteria contemplated in the different country strategies. The values of the variables were calculated considering the number of times that countries mentioned these criteria in the strategies for each school year.

Sub-sections economic investment in the SFVS, SFVS coverage, days of FV deliveries and sustainability criteria were represented using maps, and the rest were represented using a scatter plot in which each point represents the mean value of a country during the study period. To assess the relationship between variables, we calculated Pearson's correlation and linear regression. In addition, mean values were calculated for all variables. The statistical software package R [28] was used for the analysis and the graphs. The international abbreviation ISO-3166-1 ALPHA-3 was used for the nomenclature of the MS.

3. Results

3.1. General Characteristics of the SFVS by School Year in the Whole of the EU

Table 1 shows the general characteristics of the SFVS across the European Union from 2009/10 to 2016/17. During the eight years of implementation, between 21 and 25 MS participated. The percentage of EU funds used by the MS for the execution of the SFVS was irregular throughout the period, at 63% of the mean. The average participation in the SFVS—both for schools that ascribed to the SFVS and for students—increased progressively from the first to the final year, reaching a maximum in 2015/16 (44% of schools and 47% of students). The average days of FV deliveries by school year ranged from 44 to 61 days, except during the initial year in which it lasted 22 days. The average duration in weeks ranged from 13 weeks in the first year to 27 weeks in the following three years. The average over the whole period was 22 weeks. The Budget was correlated with Student expenditure

(PC 0.874 *p*-value = 0.001), participation of Schools (PC 0.799 *p*-value = 0.007) and Students (PC 0.800 *p*-value = 0.006).

Table 1. General characteristics of the EU School Fruit and Vegetables Scheme from 2009/10 to 2016/17.

School Year	MS [1]	Non-Participating MS	Budget	Student Exp [2]	Participation		Duration	
	n		%EU	€	%Schl [3]	%Stud [4]	Days of FV Deliveries	Week
09/10	21	BGR; HRV; GRC; LVA	33.11	9.35	18	21	21.75	12.96
10/11	24	HRV	61.91	14.11	38	38	59.38	26.67
11/12	23	HRV; GRC	61.77	12.48	33	34	60.61	26.63
12/13	23	HRV; CYP	71.71	15.46	39	38	54.93	26.99
13/14	25		72.98	13.42	35	42	44.37	17.75
14/15	24	GRC	66.07	12.87	28	29	44.25	16.92
15/16	24	GRC	69.25	13.38	44	47	49.05	20.38
16/17	22	GRC; LUX; ROU	67.93	13.00	44	45	57.59	25.59
Total	25		63.09	13.01	35	37	48.99	21.74

[1] MS: Member State; BGR: Bulgaria; HRV: Croatia; GRC: Greece; LVA: Latvia; CYP: Cyprus; LUX: Luxembourg; ROU: Romania. [2] Student Exp: Student Expense. [3] %Schl: percentage of schools. [4] %Stud: percentage of students.

3.2. Quantity of Fruit and Vegetables Included in the SFVS by School Year in the Whole of the EU

Table 2 shows the quantity of FV distributed among students from 2009/10 to 2016/17 for all the countries participating in the SFVS. The quantity of FV distributed to participating students was not progressive, rather there were variations in the study period, reaching a maximum in the final school year (2289.20 tons). On average, each student who participated in the SFVS received between 2 and 5 kg of FV throughout the study period, distributed in portions of 122 g. The weight of each portion ranged from 50 g (in 2009/10) to 145 g (in 2013/14), and prices changed throughout the study period, shifting from an initial cost of 0.08 € in 2009/10, and 0.42 € during the following school year to a continuous decrease, reaching 0.29 € during the 2014/15 school year.

Table 2. Quantity of Fruit and Vegetables included in the SFVS from 2009/10 to 2016/17.

School Year	Quantities of FV Purchased/Distributed	FV Per Student/School Year	Average Portions Offered per Student	Average Portion Weight	Average Portion Price
	Tons	kg	n	g	€
09/10	1707.9	1.95	15.57	49.35	0.08
10/11	631.1	4.88	51.76	139.02	0.42
11/12	1358.7	4.18	42.80	135.97	0.41
12/13	1083.1	5.54	54.23	139.03	0.40
13/14	1025.7	4.57	44.14	144.64	0.37
14/15	408.0	4.81	39.72	125.92	0.29
15/16	1743.0	3.89	41.23	120.63	0.31
16/17	2289.2	4.10	40.67	124.36	0.37
Average	1280.84	4.24	41.27	122.37	0.33

FV = fruit and vegetables.

3.3. Involvement of the MS in Carrying Out the SFVS

3.3.1. Economic Investment in the SFVS by MS during the Study Period

Figure 1 shows the percentage of EU funds used for the implementation by each MS and the expenditure per student. Four of the MS (SVK, MLT, HUN, and CZE) used 100% of EU funds to carry out the SFVS, while two (FRA and PRT) used just 25% of these funds during the period. There was important variation in the Netherlands and Austria.

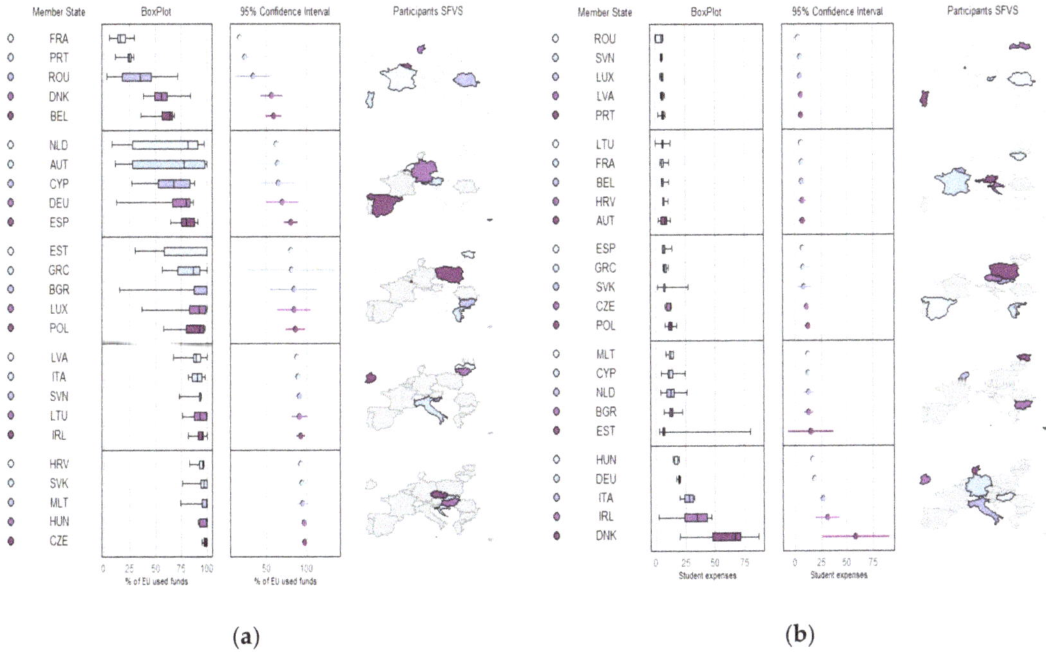

Figure 1. (**a**) Percentage of EU funds used by each MS; (**b**) expenditure per student in each MS.

In terms of expenditure per student, three countries (ITA, IRL, DNK) invested more than 25 € per student, although two of them (IRL, DNK) showed important variation during the period. In contrast, at least five countries (ROU, SVN, LUX, LVA, PRT) spent less than 6.5 € per student. Estonia stands out for its great variation during the period.

3.3.2. SFVS Coverage by MS during the Study Period

Figure 2 shows the percentage of students covered by FV distribution and EM in each MS. Two countries (FRA, DNK) barely reached 10% of students, while five (HRV, HUN, CZE, LVA, and MLT) reached between 80% and 100% of students during the period. Five other countries (ROU, CYP, AUT, BGR, LTU) stand out for the great variation in the number of students reached (between 25% and 100%) throughout the study period.

In terms of the percentage of students covered by EM, 7 of the 25 MS (BGR, DEU, GRC, ITA, LVA, LUX, NLD) were able to cover all the students registered in the SFVS with EM in all the editions. However, three countries (DNK, AUT, BEL) did not include them and did not provide data. In the rest of the MS, the percentage of students covered ranged from 15% to 90%, with great variation in most cases.

3.3.3. Days of FV Deliveries and Quantity of FV per Student (kg) by MS during the Study Period

Figure 3 shows the days of FV deliveries and quantity of FV per student (kg) in each MS. Three countries (DEU, ROU, DNK) presented a variation of 75 to 200 days during the period, while the rest did not reach 50 days. Three countries (AUT, CZE, GRC) did not report sufficient data for this variable.

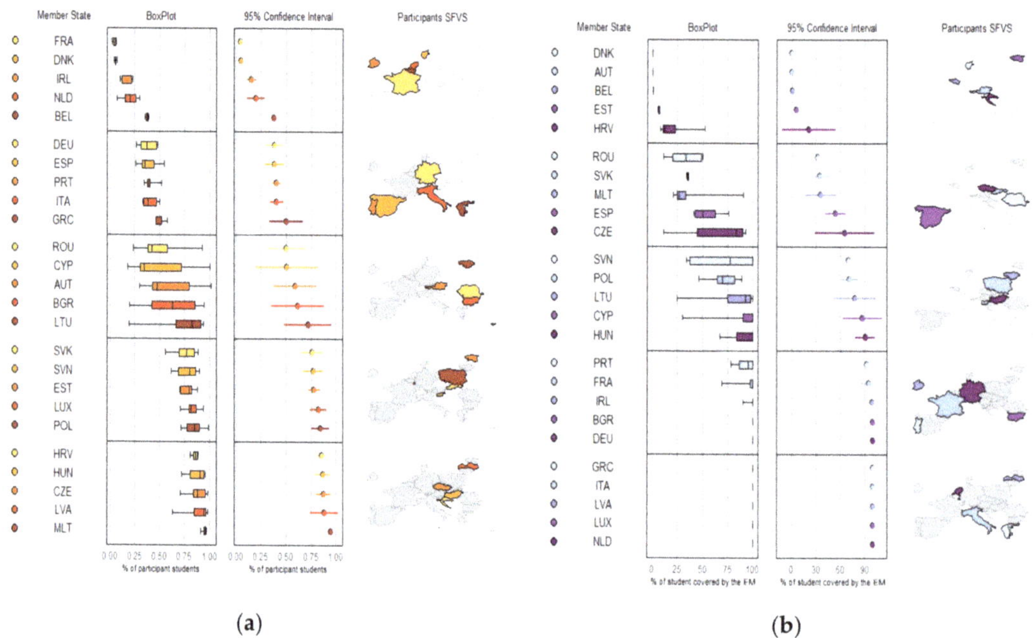

(**a**)   (**b**)

Figure 2. (**a**) Percentage of students covered by FV distribution; (**b**) Percentage of students covered by EM.

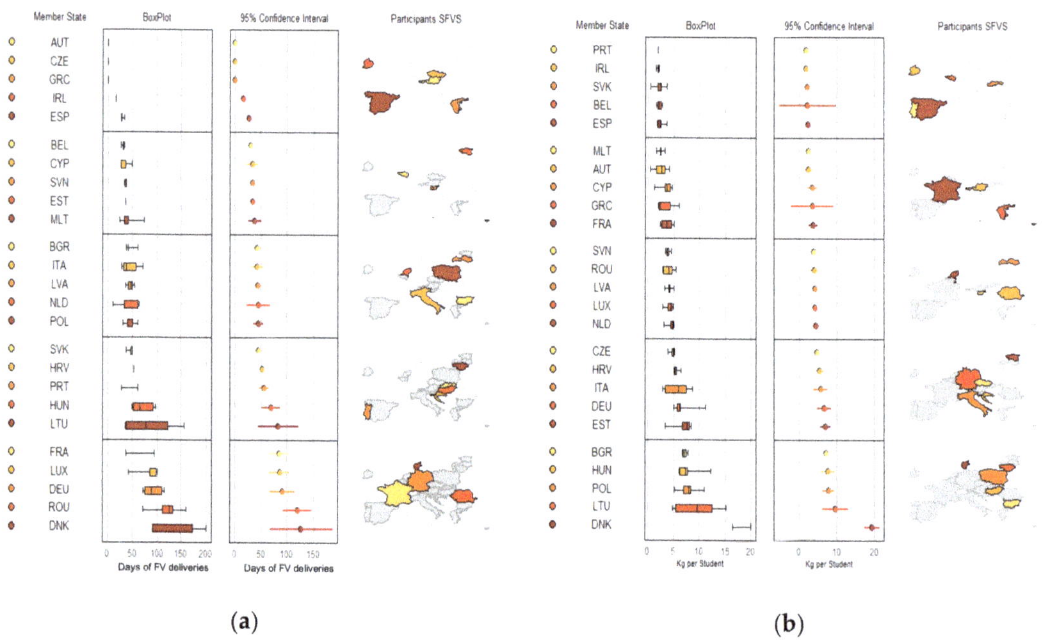

(**a**)   (**b**)

Figure 3. (**a**) Days of FV deliveries; (**b**) quantity of FV per student in each MS (kg).

The quantity of FV distributed per student varied between 2.5 and 20 kg in the countries. Six MS (PRT, IRL, SVK, BEL, ESP, and MLT) distributed around 2.5 kg of FV during the whole period, while four (BGR, HUN, POL, LTU) distributed between 7 and 15 kg per student. Denmark reached a level of between 15 and 20 kg throughout the period.

3.3.4. Economic Investment in Terms of the Coverage, Duration, and Quantity of FV Distributed by the SFVS

Figure 4 represents the relationship between the percentage of EU funds used by the MS and the percentage of students covered by the SFVS (Pearson correlation 0.637, *p*-value < 0.001). As the percentage of EU funds used increases, the percentage of students covered by the SFVS also increases. The MS that used between 75% and 100% of EU funds reached a high level of coverage (above 75%), but they distributed lower levels of FV. The MS located below the regression line did not reach the percentage of students corresponding to the use of EU funds.

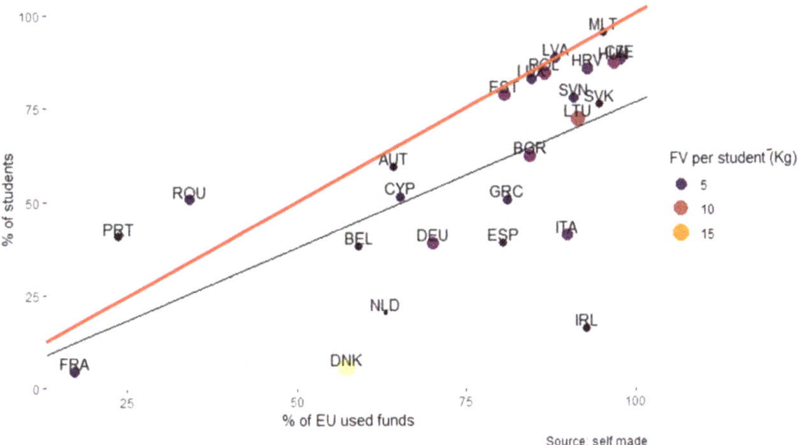

Figure 4. Relationship between the EU funds used (%), students covered (%), and the quantity of FV distributed (kg) by the SFVS. Regression line (black): represents the percentage of student beneficiaries of the SFVS using 100% of EU funds. Identity function (red): % of EU funds used by MS = % of students covered. The size of the dot represents the quantity of FV distributed per student (kg).

Figure 5 shows the relationship between the days of FV deliveries, the percentage of EU funds used by the MS (Pearson correlation −0.440, *p*-value = 0.041), and the expenditure per student (€). It can be observed that, in most cases that used more than 75% of the budget, the days of FV deliveries did not reach 50, and the expenditure per student was irregular.

Regarding the relationship between the quantity of FV distributed and the expenditure per student (Pearson Correlation 0.675, *p*-value < 0.001) and duration (Figure 6), it can be observed that most MS distributed between 4 and 8 kg of FV with an expenditure of up to 20 euros per student, and highly variable duration.

Figure 7 shows the relationship between expenditure per student (€), percent of students covered (Pearson Correlation 0.416, *p*-value = 0.043), and days of FV deliveries. It can be observed that the majority of participating MS maintained a similar expenditure per student (of approximately 10 €); however, the percentage of students covered by EM and the days of FV deliveries were more variable. A large group of MS managed to cover between 75% and 100% of students with EM, but there was a heterogeneous duration. Belgium and Austria did not include EM or did not report data in this area.

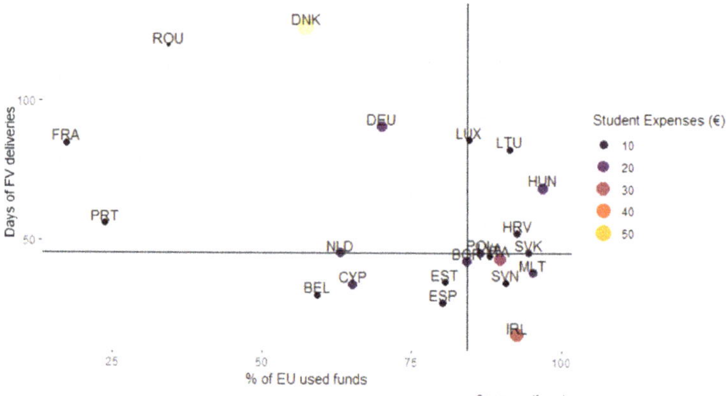

Figure 5. Relationship between the days of FV deliveries (days), the percentage of EU funds used by the MS, and the expenditure per student (€) of the SFVS. Horizontal line: represents the average of days of FV deliveries in each MS; Vertical line: represents the average % of EU funds used by the MS.

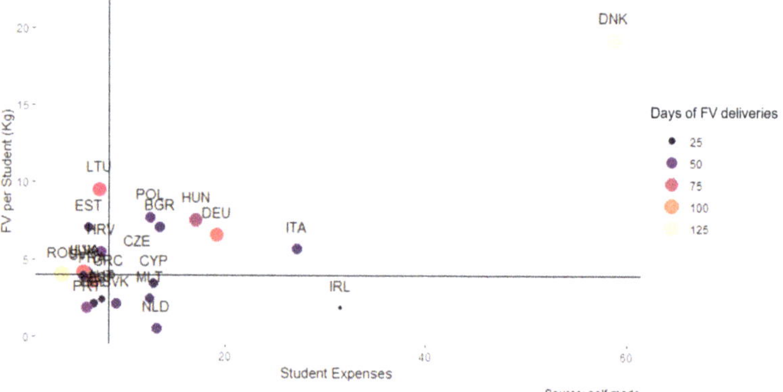

Figure 6. Relationship between fruit and vegetables distributed per student (kg), expenditure per student (€), and days of FV deliveries. Horizontal design: median FV distributed per student (3.98 kg); Vertical line: median expenditure per student (8.38 €).

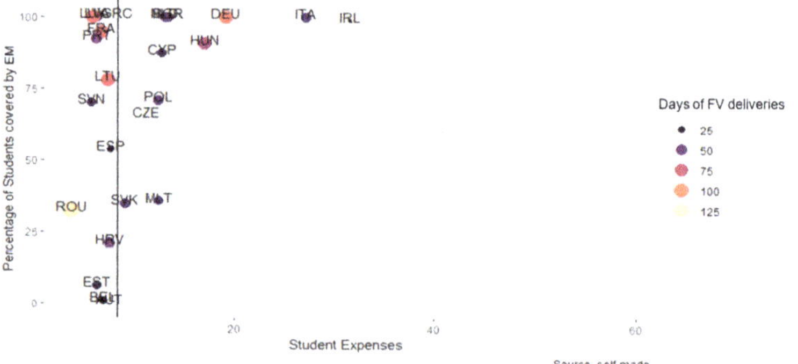

Figure 7. Relationship between the expenditure per student (€), the percentage of students covered by EM, and the days of FV deliveries. Vertical line: average expenditure per student (8.38 €).

3.4. Characteristics of FV Distribution among the Students

3.4.1. Frequency of FV Deliveries

Figure 8 shows the frequency of FV deliveries in each MS accounting SFVS duration in weeks and days of FV deliveries. Most MS designed schemes of between 2 and 3 days per week, with differences in the number of weeks of duration during the period (from 15 weeks to 35 weeks, approximately). Ireland stands out for making deliveries during one week of the school year, and Romania and Denmark stand out for distributing FV five days per week for approximately 25 weeks.

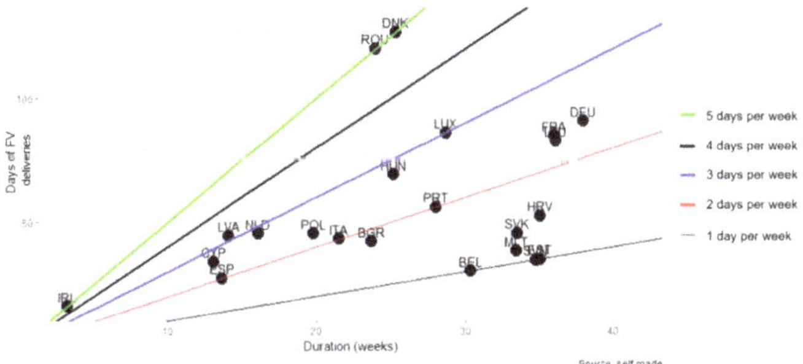

Figure 8. Relationship between SFVS duration in weeks and days of FV deliveries.

3.4.2. Quantities of FV Distributed and Percentage of Students Covered by Duration

Figure 9 shows the relationship between days of FV deliveries, quantity per student, and the percentage of students covered by the SFVS. It can be observed that the majority of MS implemented the scheme for between 25 and 50 days, with a distribution of less than 10 kg per student. MS located below the line distributed less FV than what would correspond to them, according to the programmed duration. Ireland stood out for distributing less FV for fewer days and for serving less than 25% of students during the period.

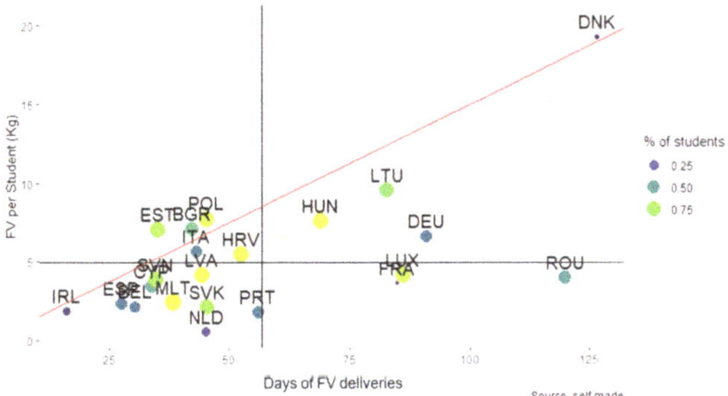

Figure 9. Relationship between days of FV deliveries, the quantity of FV per student, and the percentage of students covered by the SFVS. Horizontal line: average quantity of FV per student (4.95 kg); Vertical line: average days of FV deliveries (56.9 days). Red line: a linear equation that represents the supply of 150 g per day, based on the duration of the SFVS.

3.4.3. Portions Distributed per Student

Figure 10 shows the linear relationship between price and weight of portions. The size of the dots represents the number of portions supplied per student. It can be observed that as weight increases, the cost increases proportionally. However, all of the countries, except five, supplied portions of insufficient weight concerning cost. Except for Denmark, none of the countries surpassed 100 portions per student, and the average portion weight was less than 125 g.

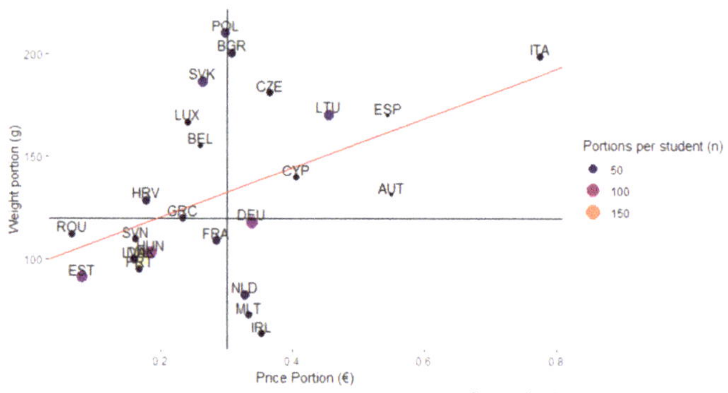

Figure 10. Relationship between portion price, portion weight, and the number of portions distributed per student by the SFVS. Horizontal line: minimum portion weight (120 g); Vertical line: average portion price (0.30 €). Regression line (red): represents the average portion price by weight.

3.4.4. Sustainability Criteria Applied in FV Distribution

Figure 11 represents the application of sustainability criteria (local, seasonal, organic, and sourcing from the European Community) for the FV distributed during the 2009/10–2016/17 period, by participating MS. In general, the most applied criteria related to seasonal produce, and organic production was the least applied criteria. Three countries (ESP, FRA, DEU) applied the four criteria for the whole period, while two (DNK and ROU) did not apply any criteria.

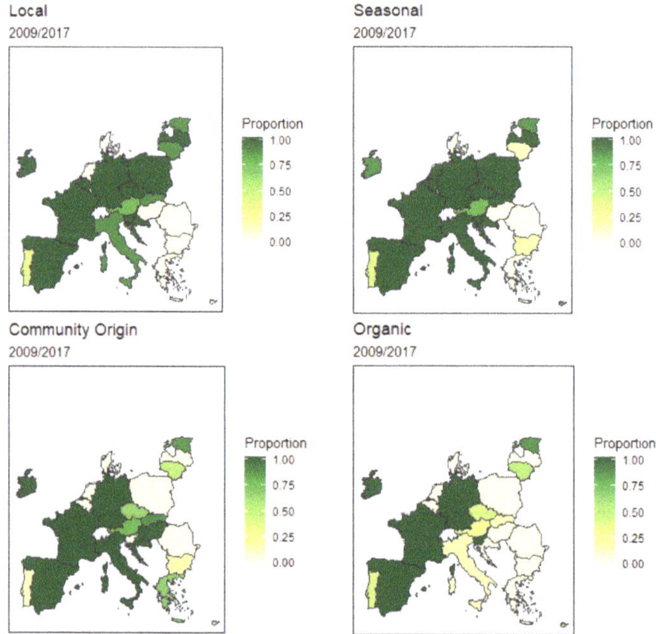

Figure 11. Sustainability criteria of the SFVS from 2009/10 to 2016/17. Average local criteria; Average temporality criteria; Average criteria European Community origin. Average organic criteria.

4. Discussion

This study identified characteristics of the implementation of the School Fruit and Vegetables Scheme from 2009/10 to 2016/17 in the European Union. The majority of MS participated in the SFVS during the study period, with a heterogeneous implementation pattern in terms of funds used, coverage, duration, quantity (totals and by portion), and cost of FV distributed per student. This heterogeneity shows different involvement of the MS that implemented the SFVS. Few countries used the entire amount of EU funds destined for SFVS implementation, made sufficient investment per student, or reached total coverage, both in terms of FV provision as well as EM. In general, the days of FV deliveries did not reach 50, and quantities of FV distributed were less than 5 kg per student. The distribution of FV also showed different characteristics between the MS. The range of periodicities of FV distributions to schools was wide, with a variation among countries of between 1 and 5 days in periods of between 1 and 35 weeks. In general, the average portion weight provided to schools was less than 125 g per day, and no country surpassed 100 portions. Temporality was the sustainability criteria most applied in the distribution of FV by different MS.

In all of the EU, the participation of schools and students in the SFVS increased across the period, however, the heterogeneity of EU funds utilized by MS had implications for the coverage reached both in terms of FV distribution as well as EM. In general, in all of the EU during the study period, the funds destined for the implementation of the SFVS seem to have been insufficient to cover the total number of students. The MS that most approximated maximum coverage quotas did so supplying little FV. In fact, only four of twenty-five MS covered the total number of students with FV, and only seven included EM in all the editions. These results are similar to those identified in the USA for the Fresh Fruit and Vegetable Program (FFVP), where coverage of 100% of the students was not achieved [29]. There could also be disadvantages for students who were not covered by the SFVS, given their exclusion from potential benefits. On one hand, there is evidence that the availability of FV in schools promotes their consumption [30,31]. On the other hand, interventions that include EM along with FV distribution are more effective in promoting the consumption of these foods in schools [32–34]. Furthermore, the educational measures contemplated in the SFVS could contribute to students' knowledge of individual health and environmental benefits related to FV consumption [35,36]. However, it is important to note that implementation of these measures falls on teachers, thus it is dependent on their willingness and perception of the additional effort involved. Providing training and support could facilitate the implementation of these measures in schools [37,38].

Despite the percentage of EU funds used permitting greater coverage of students in those MS that made a higher investment in the development of the SFVS, in general, it did not contribute to the duration nor the expenditure per student. In most cases using more than 75 percent of EU funds, the days of FV deliveries did not reach 50 days. One of the most commonly tested techniques for increasing FV consumption is repeated exposure [39]. An estimated 59 and 66 days are needed for successful habit formation [40,41], which is greater than the average number of days of provision of FV to students in the SFVS evaluated. While some countries with an expenditure per student less than the average and a long-duration plan distributed greater quantities of FV than other countries, only Denmark surpassed 15 k and had a long-duration, with significant expenditure, but with a limited reach in terms of the total number of students.

Nor did students covered by the SFVS receive enough FV constantly. As shown by the results, the quantities distributed had variation over time, providing to each school (with some exceptions) less than 5 kg per school year. These data are similar to the findings in the Polish evaluation [38]. It is known that the direct provision of FV in the school setting can have success in increasing FV consumption [10,22]. For this reason, there should be increased efforts to increase the quantity of FV distributed to students. While it is true that distribution can be influenced by different factors, plans and programs that are coordinated among different sectors could help to build alliances with all the involved parties that could contribute to achieving common objectives [42,43]. According to our results, the average portion weight was less than 125 g, and there was great variability between portion cost and portion weight among the countries. This suggests that some MS have difficulty providing the minimum portion weight of FV, which affects the consumption of the 400 g daily consumption recommended by the WHO [44,45]. Furthermore, the high heterogeneity by country in terms of FV deliveries, ranging from once per day during one week to 2.5 times per day over 35 weeks, could influence consumption habits at schools. In Germany, SFVS with FV deliveries three or two times a week led to a significant increase in FV intake [46]. For this reason, establishing a minimum quantity of FV per student and the periodicity of deliveries is recommended.

According to the SDG, it would be helpful if FV distributed in schools could contribute to reducing the environmental footprint, however, sustainability criteria were not applied uniformly in all the MS. While some countries distributed local, seasonal, and organic foods or sourced from the European Community during the study period, others did not do so in any of the editions of the SFVS. It should be noted that economic policies have promoted the importation of cheap FV to the detriment of varieties produced locally [4]. Given the limited budget, the scheme may not have a significant direct impact on market balance. Additionally, the distribution of funds to MS should be accompanied by sustainability commitments to favor uniformity of the characteristics of FV distribution in different countries. Stronger ties are needed among the different agents involved in the whole food system to integrate sustainable practices in terms of production, harvest, processing, and consumption [4].

Limitations: When interpreting the results, it should be taken into account that the data used are reported by the participating MS themselves, which could be considered a limitation of the study. However, the information reported by MS was uniform, which permitted establishing comparisons in terms of SFVS implementation in all of them. Moreover, we can measure the success of the SFVS in terms of participation, duration, or FV distributed but not in terms of the evolution of overweight and obesity, FV intake, or knowledge acquired with the educational measures because this information is not available.

5. Conclusions

Our results suggest that the implementation of the SFVS in MS has been very heterogeneous, which means that EU students do not benefit equally from the SFVS. The purchase of fruit and vegetables from local producers for distribution in schools could have a positive impact on agricultural production and also on the consumption of these products among the school population, as shown in prior studies. However, the lack of continuity in the execution, as well as the low number of days of FV deliveries, could limit its potential benefits. Establishing minimum recommendations for SFVS implementation, including the number of days, percentage of students covered, and quantities of FV distributed, are recommended to maximize the benefits of the new School Fruit, Vegetables and Milk Scheme.

Supplementary Materials: The following supporting information can be downloaded at: https://www.mdpi.com/article/10.3390/nu14153069/s1, Table S1: Database compiled from the reports in Figures 4–8. Data Base S1: Raw data for the school year 2013 with all variables used for all participating MS; Document S1: Example of annex IV: summary report 'strategy' 2014–2015 for Spain.

Author Contributions: Conceptualization, I.C.; methodology, P.C. and I.C.; software, P.C.; formal analysis, P.C.; investigation, I.C., M.A.M.-M. and P.S.; resources, I.C.; data curation, I.C. and P.C.; writing—original draft preparation, I.C.; writing—review and editing, M.C.D.-B. and P.S.; visualization, M.C.D.-B. and M.A.M.-M.; supervision, M.C.D.-B. and P.C. All authors have read and agreed to the published version of the manuscript.

Funding: This research received no external funding.

Institutional Review Board Statement: Not applicable.

Informed Consent Statement: Not applicable.

Data Availability Statement: Data supporting reported results were obtained between 2017 and 2018 from the EC web page, using the following link: https://ec.europa.eu/agriculture/sfs_en (accessed on 28 December 2017). However, this link is not currently available because the scheme changed with the entry into force of Regulation (EU) 2017/39 which lays down the basis for the new School Fruit, Vegetables and Milk Scheme. For this reason, the authors add to the Supplementary Material an example of a report from a member state and a database with all available variables for all countries for the year 2013. In addition, the authors provide the complete database upon request to the corresponding author.

Acknowledgments: Iris Comino thanks the rest of the team for their commitment to this article, which will form part of the doctoral thesis.

Conflicts of Interest: The authors declare no conflict of interest.

References

1. Afshin, A.; Sur, P.J.; Fay, K.A.; Cornaby, L.; Ferrara, G.; Salama, J.S.; Mullany, E.C.; Abate, K.H.; Abbafati, C.; Abebe, Z.; et al. Health Effects of Dietary Risks in 195 Countries, 1990–2017: A Systematic Analysis for the Global Burden of Disease Study 2017. *Lancet* **2019**, *393*, 1958–1972. [CrossRef]
2. Lim, S.; Vos, T.; Flaxman, A.; Danaei, G.; Shibuya, K.; Adair-Rohani, H.; Amann, M.; Anderson, H.; Andrews, K.; Aryee, M.; et al. A Comparative Risk Assessment of Burden of Disease and Injury Attributable to 67 Risk Factors and Risk Factor Clusters in 21 Regions, 1990–2010: A Systematic Analysis for the Global Burden of Disease Study 2010. *Lancet* **2012**, *380*, 2224–2260. [CrossRef]
3. WHO. Increasing Fruit and Vegetable Consumption to- Reduce the Risk of Noncommunicable Diseases. 2019. Available online: www.who.int/elena/titles/fruit_vegetables_ncds/en/# (accessed on 30 May 2019).
4. FAO. *Fruit and Vegetables—Your Dietary Essentials. The International Year of Fruits and Vegetables, 2021*; FAO: Rome, Italy, 2020.
5. Lynch, C.; Kristjansdottir, A.G.; Te Velde, S.J.; Lien, N.; Roos, E.; Thorsdottir, I.; Krawinkel, M.; De Almeida, M.D.V.; Papadaki, A.; Hlastan Ribic, C.; et al. Fruit and Vegetable Consumption in a Sample of 11-Year-Old Children in Ten European Countries–the PRO GREENS Cross-Sectional Survey. *Public Health Nutr.* **2014**, *17*, 2436–2444. [CrossRef] [PubMed]
6. OECD/European Union. Fruit and Vegetable Consumption among Children. In *Health at a Glance: Europe 2016: State of Health in the EU Cycle*; OECD Publishing: Paris, France, 2016; pp. 100–101.
7. Williams, J.; Buoncristiano, M.; Nardone, P.; Rito, A.I.; Spinelli, A.; Hejgaard, T.; Kierkegaard, L.; Nurk, E.; Kunešová, M.; Milanović, S.M.; et al. A Snapshot of European Children's Eating Habits: Results from the Fourth Round of the WHO European Childhood Obesity Surveillance Initiative (COSI). *Nutrients* **2020**, *12*, 2481. [CrossRef] [PubMed]
8. WHO. Growing up Unequal: Gender and Socioeconomic Differences in Young People's Health and Well-Being. 2016. Available online: https://apps.who.int/iris/handle/10665/326320 (accessed on 30 May 2019).
9. Evans, G.W.; Jones-Rounds, M.L.; Belojevic, G.; Vermeylen, F. Family Income and Childhood Obesity in Eight European Cities: The Mediating Roles of Neighborhood Characteristics and Physical Activity. *Soc. Sci. Med.* **2012**, *75*, 477–481. [CrossRef]
10. Craigie, A.M.; Lake, A.A.; Kelly, S.A.; Adamson, A.J.; Mathers, J.C. Tracking of Obesity-Related Behaviours from Childhood to Adulthood: A Systematic Review. *Maturitas* **2011**, *70*, 266–284. [CrossRef]
11. Mikkilä, V.; Räsänen, L.; Raitakari, O.T.; Pietinen, P.; Viikari, J. Longitudinal Changes in Diet from Childhood into Adulthood with Respect to Risk of Cardiovascular Diseases: The Cardiovascular Risk in Young Finns Study. *Eur. J. Clin. Nutr.* **2004**, *58*, 1038–1045. [CrossRef]
12. Triador, L.; Farmer, A.; Maximova, K.; Willows, N.; Kootenay, J. A School Gardening and Healthy Snack Program Increased Aboriginal First Nations Children's Preferences toward Vegetables and Fruit. *J. Nutr. Educ. Behav.* **2015**, *47*, 176–180. [CrossRef]
13. Horne, P.J.; Tapper, K.; Lowe, C.F.; Hardman, C.A.; Jackson, M.C.; Woolner, J. Increasing Children's Fruit and Vegetable Consumption: A Peer-Modelling and Rewards-Based Intervention. *Eur. J. Clin. Nutr.* **2004**, *58*, 1649–1660. [CrossRef]
14. DeCosta, P.; Møller, P.; Frøst, M.B.; Olsen, A. Changing Children's Eating Behaviour—A Review of Experimental Research. *Appetite* **2017**, *113*, 327–357. [CrossRef]
15. Wilson, D.B.; Jones, R.M.; McClish, D.; Westerberg, A.L.; Danish, S. Fruit and Vegetable Intake among Rural Youth Following a School-Based Randomized Controlled Trial. *Prev. Med.* **2011**, *54*, 150–156. [CrossRef]
16. Delgado-Noguera, M.; Tort, S.; Martínez-Zapata, M.J.; Bonfill, X. Primary School Interventions to Promote Fruit and Vegetable Consumption: A Systematic Review and Meta-Analysis. *Prev. Med.* **2011**, *53*, 3–9. [CrossRef]
17. Knai, C.; Pomerleau, J.; Lock, K.; McKee, M. Getting Children to Eat More Fruit and Vegetables: A Systematic Review. *Prev. Med.* **2006**, *42*, 85–95. [CrossRef]
18. Blanchette, L.; Brug, J. Determinants of Fruit and Vegetable Consumption among 6–12-Year-Old Children and Effective Interventions to Increase Consumption. *J. Hum. Nutr. Diet.* **2005**, *18*, 431–443. [CrossRef]
19. De Sa, J.; Lock, K. Will European Agricultural Policy for School Fruit and Vegetables Improve Public Health? A Review of School Fruit and Vegetable Programmes. *Eur. J. Public Health* **2008**, *18*, 558–568. [CrossRef]

20. Krølner, R.; Rasmussen, M.; Brug, J.; Klepp, K.-I.; Wind, M.; Due, P. Determinants of Fruit and Vegetable Consumption among Children and Adolescents: A Review of the Literature. Part II: Qualitative Studies. *Int. J. Behav. Nutr. Phys. Act.* **2011**, *8*, 112. [CrossRef]

21. Van Cauwenberghe, E.; Maes, L.; Spittaels, H.; Van Lenthe, F.J.; Brug, J.; Oppert, J.M.; De Bourdeaudhuij, I. Effectiveness of School-Based Interventions in Europe to Promote Healthy Nutrition in Children and Adolescents: Systematic Review of Published and Grey Literature. *Br. J. Nutr.* **2010**, *103*, 781–797. [CrossRef]

22. Micha, R.; Karageorgou, D.; Bakogianni, I.; Trichia, E.; Whitsel, L.P.; Story, M.; Peñalvo, J.L.; Mozaffarian, D. Effectiveness of School Food Environment Policies on Children's Dietary Behaviors: A Systematic Review and Meta-Analysis. *PLoS ONE* **2018**, *13*, e0194555. [CrossRef]

23. Tak, N.I.; Te Velde, S.J.; Brug, J. Long-Term Effects of the Dutch Schoolgruiten Project—Promoting Fruit and Vegetable Consumption among Primary-School Children. *Public Health Nutr.* **2009**, *12*, 1213–1223. [CrossRef]

24. Rosi, A.; Scazzina, F.; Ingrosso, L.; Morandi, A.; Del Rio, D.; Sanna, A. The '5 a Day' Game: A Nutritional Intervention Utilising Innovative Methodologies with Primary School Children. *Int. J. Food Sci. Nutr.* **2015**, *66*, 713–717. [CrossRef]

25. Council of The European Union. *Council Regulation (EC) No 1182/2007 of 26 September 2007*; Council of The European Union: Strasbourg, France, 2007.

26. Council of The European Union. *Council Regulation (EC) No 13/2009 of 18 December 2008*; Council of The European Union: Strasbourg, France, 2008.

27. Willett, W.; Rockström, J.; Loken, B.; Springmann, M.; Lang, T.; Vermeulen, S.; Garnett, T.; Tilman, D.; DeClerck, F.; Wood, A.; et al. Food in the Anthropocene: The EAT–Lancet Commission on healthy diets from sustainable food systems. *Lancet* **2019**, *393*, 447–492. [CrossRef]

28. R Core Team. *R: A Language and Environment for Statistical Computing*; R Core Team: Vienna, Austria, 2021.

29. Ohri-Vachaspati, P.; Turner, L.; Chaloupka, F.J. Fresh Fruit and Vegetable Program Participation in Elementary Schools in the United States and Availability of Fruits and Vegetables in School Lunch Meals. *J. Acad. Nutr. Diet.* **2012**, *112*, 921–926. [CrossRef]

30. Holley, C.; Farrow, C.; Haycraft, E. A Systematic Review of Methods for Increasing Vegetable Consumption in Early Childhood. *Curr. Nutr. Rep.* **2017**, *6*, 157–170. [CrossRef]

31. Sirasa, F.; Mitchell, L.; Rigby, R.; Harris, N. Family and Community Factors Shaping the Eating Behaviour of Preschool-Aged Children in Low and Middle-Income Countries: A Systematic Review of Interventions. *Prev. Med.* **2019**, *129*, 105827. [CrossRef]

32. Hodder, R.K.; O'Brien, K.M.; Tzelepis, F.; Wyse, R.J.; Wolfenden, L. Interventions for Increasing Fruit and Vegetable Consumption in Children Aged Five Years and Under. *Cochrane Database Syst. Rev.* **2020**, *5*, CD008552. [CrossRef]

33. Singhal, J.; Herd, C.; Adab, P.; Pallan, M. Effectiveness of School-Based Interventions to Prevent Obesity among Children Aged 4 to 12 Years Old in Middle-Income Countries: A Systematic Review and Meta-Analysis. *Obes. Rev.* **2021**, *22*, e13105. [CrossRef]

34. Mak, T.; Storcksdieck Genannt Bonsmann, S.; Louro Caldeira, S.; Wollgast, J. *How to Promote Fruit and Vegetable Consumption in Schools: A Toolkit, EUR 27946*; Publications Office of the European Union: Luxembourg, 2016. [CrossRef]

35. Fink, L.; Strassner, C.; Ploeger, A. Exploring External Factors Affecting the Intention-Behavior Gap When Trying to Adopt a Sustainable Diet: A Think Aloud Study. *Front. Nutr.* **2021**, *8*, 511412. [CrossRef]

36. Vermeir, I.; Verbeke, W. Sustainable Food Consumption: Exploring the Consumer "Attitude—Behavioral Intention" Gap. *J. Agric. Environ. Ethics* **2006**, *19*, 169–194. [CrossRef]

37. Roccaldo, R.; Censi, L.; D'Addezio, L.; Berni Canani, S.; Gennaro, L. A Teachers' Training Program Accompanying the "School Fruit Scheme" Fruit Distribution Improves Children's Adherence to the Mediterranean Diet: An Italian Trial. *Int. J. Food Sci. Nutr.* **2017**, *68*, 887–900. [CrossRef]

38. Wolnicka, K.; Taraszewska, A.M.; Jaczewska-Schuetz, J. Can the School Fruit and Vegetable Scheme Be an Effective Strategy Leading to Positive Changes in Children's Eating Behaviours? Polish Evaluation Results. *Int. J. Environ. Res. Public Health* **2021**, *18*, 12331. [CrossRef]

39. Appleton, K.M.; Hemingway, A.; Rajska, J.; Hartwell, H. Repeated Exposure and Conditioning Strategies for Increasing Vegetable Liking and Intake: Systematic Review and Meta-Analyses of the Published Literature. *Am. J. Clin. Nutr.* **2018**, *108*, 842. [CrossRef] [PubMed]

40. Keller, J.; Kwasnicka, D.; Klaiber, P.; Sichert, L.; Lally, P.; Fleig, L. Habit Formation Following Routine-Based versus Time-Based Cue Planning: A Randomized Controlled Trial. *Br. J. Health Psychol.* **2021**, *26*, 807–824. [CrossRef] [PubMed]

41. Lally, P.; van Jaarsveld, C.H.M.; Potts, H.W.W.; Wardle, J. How Are Habits Formed: Modelling Habit Formation in the Real World. *Eur. J. Soc. Psychol.* **2010**, *40*, 998–1009. [CrossRef]

42. Swinburn, B.; Kraak, V.; Rutter, H.; Vandevijvere, S.; Lobstein, T.; Sacks, G.; Gomes, F.; Marsh, T.; Magnusson, R. Strengthening of Accountability Systems to Create Healthy Food Environments and Reduce Global Obesity. *Lancet* **2015**, *385*, 2534–2545. [CrossRef]

43. Baker, P.; Hawkes, C.; Wingrove, K.; Demaio, A.R.; Parkhurst, J.; Thow, A.M.; Walls, H. What Drives Political Commitment for Nutrition? A Review and Framework Synthesis to Inform the United Nations Decade of Action on Nutrition. *BMJ Glob. Health* **2018**, *3*, 485. [CrossRef]

44. Mason-D'Croz, D.; Bogard, J.; Sulser, T.; Cenacchi, N.; Dunston, S.; Herrero, M.; Wiebe, K. Gaps between Fruit and Vegetable Production, Demand, and Recommended Consumption at Global and National Levels: An Integrated Modelling Study. *Lancet Planet. Health* **2019**, *3*, e318–e329. [CrossRef]

45. WHO. *Diet, Nutrition and the Prevention of Chronic Diseases: Report of a Joint WHO/FAO Expert Consultation*; WHO: Geneva, Switzerland, 2003.
46. Haß, J.; Lischetzke, T.; Hartmann, M. Does the Distribution Frequency Matter? A Subgroup Specific Analysis of the Effectiveness of the EU School Fruit and Vegetable Scheme in Germany Comparing Twice and Thrice Weekly Deliveries. *Public Health Nutr.* **2018**, *21*, 1375–1387. [CrossRef]

 *nutrients*

Article

Evidence Use in the Development of the Australian Dietary Guidelines: A Qualitative Study

Kate Wingrove *, Mark A. Lawrence, Cherie Russell and Sarah A. McNaughton

Institute for Physical Activity and Nutrition (IPAN), School of Exercise and Nutrition Sciences, Deakin University, Geelong 3220, Australia; mark.lawrence@deakin.edu.au (M.A.L.); caru@deakin.edu.au (C.R.); sarah.mcnaughton@deakin.edu.au (S.A.M.)
* Correspondence: k.wingrove@deakin.edu.au

Abstract: Dietary guidelines are important nutrition policy reference standards that should be informed by the best available evidence. The types of evidence that are reviewed and the evidence review methods that are used have implications for evidence translation. The aim of this study was to explore perceived advantages, disadvantages, and practicalities associated with the synthesis and translation of evidence from nutrient-based, food-based, and dietary patterns research in dietary guideline development. A qualitative descriptive study was conducted. Twenty-two semi-structured interviews were conducted with people involved in the development of the 2013 Australian Dietary Guidelines (ADGs). Transcripts were analysed thematically. To inform future ADGs, there was support for reviewing evidence on a range of dietary exposures (including dietary patterns, foods and food groups, nutrients and food components, and eating occasions) and health outcomes, as well as evidence on environmental sustainability and equity. At the evidence synthesis stage, practicalities associated with planning the evidence review and conducting original systematic reviews were discussed. At the evidence translation stage, practicalities associated with integrating the evidence and consulting stakeholders were described. To ensure that the best available evidence is translated into future ADGs, evidence review methods should be selected based on the exposures and outcomes of interest.

Keywords: dietary guidelines; dietary patterns; evidence synthesis; evidence translation; qualitative research

Citation: Wingrove, K.; Lawrence, M.A.; Russell, C.; McNaughton, S.A. Evidence Use in the Development of the Australian Dietary Guidelines: A Qualitative Study. *Nutrients* **2021**, *13*, 3748. https://doi.org/10.3390/nu13113748

Academic Editor: George Moschonis

Received: 17 September 2021
Accepted: 21 October 2021
Published: 23 October 2021

Publisher's Note: MDPI stays neutral with regard to jurisdictional claims in published maps and institutional affiliations.

1. Introduction

Dietary guidelines provide recommendations on the foods and dietary patterns that are associated with reduced diet-related chronic disease and obesity risk and provide sufficient amounts of the nutrients required to promote health [1–3]. Dietary guideline statements are often accompanied by food guides that are designed to provide practical advice on selecting quantities, combinations, and varieties of foods to achieve healthy dietary patterns [3]. Dietary guidelines can be used to underpin policies and programs across a range of sectors, including health, education, and agriculture [3,4]. They can also be used by health professionals who work directly with members of the public as nutrition education tools [3,4].

The FAO provides guidance to member states on the development of dietary guidelines. The dietary guideline development process typically involves the establishment of an expert committee that is responsible for synthesising and translating evidence into dietary guidelines [1–3]. Since the FAO report on the preparation and use of food-based dietary guidelines was published in 1998, perspectives on the types of evidence that should be reviewed and the evidence review methods that should be used have evolved [1,5,6]. The types of evidence that are reviewed and the methods that are used to review the evidence have implications for evidence translation [6–8].

According to the FAO, dietary guidelines should be informed by a review of the best available evidence on associations between diet and health [1,2]. Since the emergence of the field of nutrition science in the 1900s, associations between dietary exposures (e.g., nutrients, foods, and dietary patterns) and health outcomes (e.g., micronutrient status, chronic disease risk factors, chronic disease incidence, and mortality) have been explored [9,10]. Dietary patterns are defined by the USDA as the "quantities, proportions, variety, or combination of different foods, drinks, and nutrients in diets and the frequency with which they are habitually consumed" [11] (p. 8). In the last 20 years, the volume of evidence on associations between dietary patterns and health outcomes has increased substantially [12,13]. This evidence is derived primarily from prospective cohort studies that have been conducted over years or decades [14–16] examining the associations between dietary patterns and long-term health outcomes, including chronic disease incidence and mortality [17–19]. The need to review evidence from dietary patterns research alongside evidence from food-based and nutrient-based research to inform dietary guidelines is now well accepted [7,9,13].

It is now expected that dietary guidelines are underpinned by systematic reviews that are conducted in line with best practice guidelines [5,6,8]. The WHO and Cochrane describe the following steps: define the research question in terms of the Population, Intervention (or exposure), Comparator, and Outcome (PICO) of interest; develop inclusion criteria that reflect the research question; identify studies that meet the inclusion criteria; extract data from included studies; assess the risk of bias associated with each included study; synthesise the data from included studies (using a meta-analysis where possible); assess the quality (or certainty) of the body of evidence for each outcome using the Grading of Recommendations, Assessment, Development and Evaluation (GRADE) system; and interpret the results and draw conclusions [20,21]. In recent years, concerns have been raised that systematic review methods that were developed for other purposes may not be appropriate for use in dietary guideline development [8,22]. For example, the GRADE approach was designed to assess the quality of evidence from randomised controlled trials (RCTs) and non-randomised intervention studies [23]. Depending on the research questions that are asked, approaches that have been designed for the purpose of assessing the quality of evidence from observational studies with complex exposures and long-term health outcomes may be more appropriate [22,24].

At the evidence translation stage, the expert committee considers the quality of the evidence that has been reviewed alongside contextual factors including the social, economic, and political environment [1–3]. The dietary guidelines are then drafted, and stakeholders are consulted. As part of the consultation process, dietary guideline statements, food guides, and other resources are tested with consumers (including health professionals and members of the general public) [1–3]. The dietary guidelines are then finalised and disseminated [1,2].

The Dietary Guidelines for Australians were first published by the Department of Health in 1982 [25]. Revised guidelines were published in 1992 [26], followed by the publication of specific guidelines for children and adolescents in 1995 [27] and for older adults in 1999 [28]. Revised guidelines were published in 2003 for children and adolescents [29] and for adults [30]. The current Australian Dietary Guidelines (ADGs) were published in 2013 and include five dietary guideline statements and the Australian Guide to Healthy Eating (AGHE) [31,32]. The ADGs were informed by a combination of original systematic reviews and narrative reviews. Our previous analysis demonstrated that most of the systematic reviews on diet and health synthesised evidence from food-based research, while only a small proportion synthesised evidence from dietary patterns research [33]. In July 2020, a review of the ADGs was announced [34]. In developing the next iteration of the ADGs, there is an opportunity to review the latest evidence on associations between dietary patterns and health outcomes. However, the use of systematic review methods that do not take into consideration the nature of dietary patterns evidence may influence evidence translation [7,35]. A description of the challenges associated with conducting the systematic reviews that informed the 2013 ADGs was published in 2014 [36]. A combination of methodological challenges (e.g., accurate assessment of the quality of evidence from prospective cohort studies with dietary exposures) and practical challenges (e.g., the resource-intensive nature of the systematic review process) were described [36]. The aim of this study was to explore perceived advantages, disadvantages, and practicalities associated with the synthesis and translation of evidence from nutrient-based, food-based, and dietary patterns research in dietary guideline development.

2. Materials and Methods

A qualitative descriptive study design was used to answer questions about evidence use in dietary guideline development [37,38]. The lead researcher (KW) held a relativist ontological position and used an epistemology that embraced subjectivity [39,40]. This means that she was not looking to identify one 'absolute truth' but was instead seeking to explore the participants' experiences, ideas and opinions. The methods and results of this study have been reported according to the Consolidated criteria for reporting qualitative research (COREQ) checklist [41].

Purposive sampling methods were used to recruit participants [42]. The following people who contributed to the evidence review that informed the ADGs were eligible to participate in this study: members of the 2013 Australian Dietary Guidelines Working Committee (n = 12, excluding ML because his dual roles as a member of the research team and as a member of the Working Committee may be perceived to be a source of potential bias); and members of the Dietitians Association of Australia Review Team (n = 36). The Working Committee was appointed by the National Health and Medical Research Council (NHMRC) and included experts in nutrition and public health and representatives from food industry and consumer groups [31,43]. The Working Committee was responsible for developing the research questions and the evidence review methodology (in line with NHMRC procedures and in collaboration with methodologists), and translating the evidence into dietary guideline recommendations [31]. The Review Team was commissioned by the NHMRC to conduct the evidence review [31,44]. The Review Team included a review leadership team (three senior dietitians), a project officer, a project manager, and 29 reviewers (all dietitians), and two subject librarians [44]. The names and email addresses of eligible participants were identified using publicly available information. A plain language statement and consent form was emailed to eligible participants at the time of recruitment. Informed consent was obtained in writing prior to the commencement of each interview. The concept of information power can be used to determine the sample size for qualitative studies based on the relevance of the information provided by participants [45,46]. Due to the small number of people with the experience required to participate in this study, the sample size was not determined based on the concept of information power [45,46]. Instead, recruitment ceased when no further responses were received from eligible participants. Participants received a $20 WISH eGift Card as compensation for their time.

Semi-structured interviews were conducted online (using Zoom) or over the phone by KW (an Accredited Practising Dietitian and PhD student with training and experience in conducting qualitative research). A semi-structured interview guide was developed and tested in a pilot interview with ML prior to data collection (Table S1) [47]. Participants were asked about their involvement in the dietary guideline development process, the types of evidence that were used to inform the 2013 ADGs, and the types of evidence that should be used to inform future dietary guidelines. The semi-structured nature of the interviews allowed the questions to be tailored to the experiences of each participant [39]. For example, people involved in conducting systematic reviews were asked about the practicalities associated with the evidence synthesis process, and people involved in the development of the dietary guideline statements were asked in more detail about the practicalities associated with evidence translation. Issues relating to the implementation and evaluation of dietary guidelines were beyond the scope of this study.

Throughout the data collection process, minor adaptations to the interview guide were made based on responses from interviewees (e.g., broad questions that were poorly understood were clarified, or additional prompts were added) and in response to the release of new information from the NHMRC about the ADG review process. To enhance reflexivity, KW completed a written reflection on the interview process immediately after each interview [39,42]. The reflection template included factors that may have influenced the data (e.g., variation in interview technique) and preliminary data analysis ideas. Interviews were audio-recorded and transcribed verbatim using artificial intelligence software (Otter.ai). Transcripts were checked for accuracy and edited accordingly by KW, and any information that could be used to identify a participant was removed [48]. Transcripts were not returned to participants.

An iterative thematic approach to data analysis was used, which is consistent with the qualitative descriptive research design [37]. An inductive, open coding technique was used, whereby previously undefined codes were assigned to pieces of data [39,49]. A 10% sample of interview transcripts (n = 3) was independently analysed by two researchers (KW and CR). Each researcher developed a preliminary coding framework to facilitate identification of themes in the data, and the differences between these coding frameworks were discussed [49,50]. The remaining interviews were analysed by one researcher (KW). The coding framework was adapted throughout the data analysis process until the themes provided an accurate representation of the data [37]. NVivo software (Version 12 Plus) was used to facilitate data analysis [51].

3. Results

Forty-eight people were eligible to participate in this study (Figure 1). Five people were unable to be contacted via email (email addresses were identified, but email delivery failed). Forty-three people were contacted. Twelve people did not respond, three people declined due to a lack of time, and two people declined due to a perceived lack of expertise. Twenty-six people expressed interest in participating; however, four of these people did not respond to follow-up emails. Individual semi-structured interviews were conducted with the remaining 22 people between October and December 2020. Six participants were members of the Working Committee, and 16 participants were members of the Review Team. The length of the interviews ranged from 30 to 60 min.

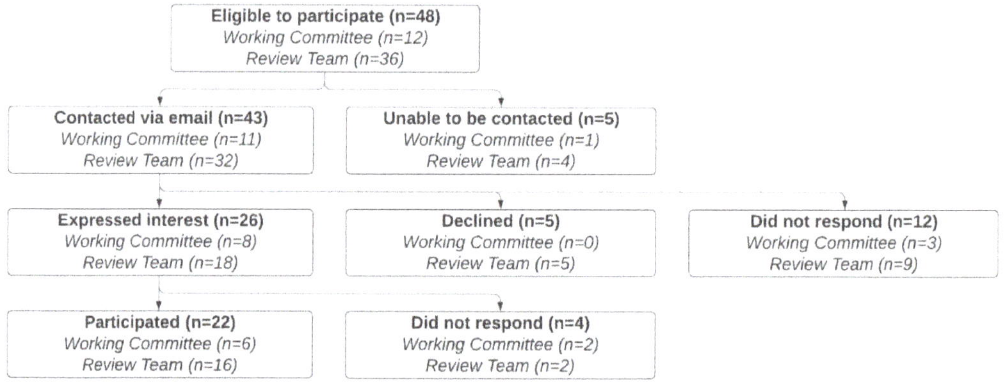

Figure 1. Recruitment and participant characteristics.

Eight themes emerged from the data and were organised under three categories (Table 1). Four themes related to exposures and outcomes of interest, two themes related to practicalities associated with evidence synthesis, and two themes related to practicalities associated with evidence translation. Some participants also shared their views on the broader contextual issues associated with dietary guideline development in Australia (such as the processes used to establish the Working Committee and the need for regular, planned revisions of the dietary guidelines). Although these are important issues, they were outside the scope of the aim of this study and, therefore, have not been described in further detail.

Table 1. Overview of the eight themes that emerged from the data, organised under three categories.

Categories		Themes
Exposures and outcomes of interest	1.	Dietary exposures
	2.	Health outcomes
	3.	Environmental sustainability
	4.	Equity
Practicalities associated with evidence synthesis	5.	Planning the evidence review
	6.	Conducting original systematic reviews
Practicalities associated with evidence translation	7	Integrating the evidence
	8.	Consulting stakeholders

3.1. Exposures and Outcomes of Interest

In relation to the evidence review that will be conducted to inform the next iteration of the ADGs, participants identified a range of dietary exposures and health outcomes of interest. The importance of reviewing the latest evidence on environmental sustainability and equity was also described. Opinions on the research questions that should be prioritised reflected participants' views on the public health nutrition issues that future dietary guidelines should aim to address, and their perspectives on the current state of knowledge on relationships between particular exposures and outcomes.

3.1.1. Dietary Exposures

The following dietary exposures of interest were identified: dietary patterns, foods and food groups, nutrients and food components, and eating occasions. Conceptually, dietary patterns, foods and food groups, and nutrients and food components are exposures that reflect dietary intake (i.e., 'what' people eat), whereas eating occasions are exposures that reflect eating behaviours (i.e., 'how' people eat) [52]. Dietary patterns and foods and food groups were consistently described as important exposures. Some participants described particular nutrients and food components as important exposures in relation to particular health outcomes. Views on the importance of reviewing evidence on eating occasions were mixed.

In line with the food synergy theory [16,18], dietary patterns were described as exposures of interest on the basis that it is the combinations of foods that appear to be most influential on important health outcomes. For example:

> " ... if we're defining health in terms of maximising functionality and preventing chronic disease, which is a public health goal, then we need to accept that it's not a single food, it's not a single nutrient, it's actually a package, which we are now calling dietary patterns, which influences those outcomes, and it doesn't happen quickly" [Participant 3, Working Committee].

Food and food groups were described as exposures of interest because although people consume dietary patterns, not every dietary pattern includes every food. For example:

> " ... I guess if you've got a dietary pattern, it might not include all the different foods. Whereas if you've got food studies, for example, we know that fatty fish is high in omega-3. Those studies are still important, because that's a strong source of that particular nutrient" [Participant 20, Review Team].

Some participants identified particular food groups of interest (e.g., meat, dairy, discretionary foods), explaining that new evidence has emerged since the current ADGs were developed.

Nutrients and food components were described as exposures of interest on the basis that we need to understand the biological mechanisms that underpin associations between particular foods and health outcomes. For example:

"... since we're talking about the biological aspects, we do need to have a good understanding of the molecular basis for how food components influence health, and that's actually the nutrient side of it. But it's not just nutrients, it's the things we don't call nutrients, like phytocomponents, and it's also the interaction between nutrients that occur within the food delivery system" [Participant 3, Working Committee].

The following nutrients and food components were described as exposures of interest in relation to particular health outcomes: different types of saturated fat in relation to blood lipids, different types of carbohydrate (e.g., sugars compared to starches) in relation to body weight, potassium (including the potassium to sodium ratio) in relation to cardiovascular disease risk, flavonoids in relation to vascular health, and food additives (including stabilisers) in relation to inflammation.

Eating occasions (including the timing and frequency of meals, as well as contextual factors such as eating with others, eating away from home, and eating home-prepared meals) were described as exposures of interest by some participants on the basis that there is emerging evidence to suggest that these exposures may be associated with a range of important health outcomes. Other participants were of the opinion that research questions about 'what' people should eat should take priority over research questions about 'how' people should eat.

3.1.2. Health Outcomes

Cardiovascular disease, type 2 diabetes, cancer, and overweight and obesity were often described as important health outcomes. The following health outcomes of interest were also identified: gut health (including the microbiome), gastrointestinal diseases, food allergies, mental health, immune function, vitamin D deficiency, healthy ageing, sarcopenia, frailty, osteoarthritis, and cognitive outcomes (including dementia and Alzheimer's disease).

3.1.3. Environmental Sustainability

There was strong support for an increased focus on environmental sustainability in the development of future dietary guidelines on the basis that it is not possible to promote public health without considering sustainability and in response to increasing environmental threats to food security in Australia. For example:

"So the first thing I'd say is that you don't even have to put those two things together. You don't even have to say health and environmental sustainability, because anything that degrades environmental sustainability eventually degrades human health, and that's becoming more and more accepted" [Participant 12, Review Team].

Some participants commented on the broad nature of the term 'environmental sustainability' and explained that there are many research questions that could be asked. Reviewing the latest Australian evidence on the environmental impacts associated with food production and consumption was identified as a priority due to an increase in the volume of relevant evidence. For example:

"We now have much more evidence than we had back then. And not just of modelling studies of actual observational data. So in terms of environment and food related health, so we've got much more data" [Participant 4, Working Committee].

Other participants explained that although environmental sustainability was important to consider, the evidence review should focus on dietary exposures and health outcomes, and environmental sustainability should be considered as part of the evidence translation process.

3.1.4. Equity

There was strong support for an increased focus on equity in the development of future dietary guidelines on the basis that access to food is a social determinant of health and in response to increasing economic threats to food security in Australia. Some participants explained that 'equity' is a broad term, and that there are many research questions that could be asked. Reviewing the latest Australian evidence on relationships between socioeconomic status and dietary intake was identified as a priority. For example:

> "So if there's anything I think we should be looking at, it's the relationship between socioeconomic status and dietary intake. I'm not a food security expert by any means, I've just been an interested observer during the pandemic, and how people's food skills, so utilisation has been poor, worse than that, we've got people economically doing it tough, not having enough to eat. So I think if there's any time to be asking questions, that's an important one to be asking of the literature" [Participant 22, Review Team].

Other participants explained that although equity issues were important to consider, the evidence review should focus on dietary exposures and health outcomes, and equity should be considered as part of the evidence translation process.

3.2. Practicalities Associated with Evidence Synthesis

In relation to the evidence synthesis stage of dietary guideline development, practicalities associated with planning the evidence review and conducting original systematic reviews were described.

3.2.1. Planning the Evidence Review

The importance of having an appropriate methodology in place before the evidence review begins was emphasised by some participants. For example:

> "So rather than say, 'go and do some dietary guidelines again', it's 'we thought about how this needs to be done, now here's the methodology'. And then it's like the rules of the game have been stipulated before everyone goes on the field; it's much more functional" [Participant 3, Working Committee].

Many participants described the importance of balancing evidence review methods with available resources, including human resources, financial resources, and time. The advantages and disadvantages of conducting original systematic reviews for the purpose of dietary guideline development (rather than using existing systematic reviews that were conducted for other purposes) were discussed. The resource-intensive nature of conducting original systematic reviews was emphasised, and there was support for using a combination of original and existing systematic reviews on this basis. If existing systematic reviews are used, the importance of assessing the quality of those reviews and assessing the applicability of the evidence to the Australian context was described. On this basis, some participants argued that the process of identifying and assessing existing systematic reviews can be less efficient than conducting original systematic reviews.

3.2.2. Conducting Original Systematic Reviews

In relation to conducting original systematic reviews, practicalities at the following stages of the systematic review process were described: defining the research question, developing the search strategy and inclusion criteria, assessing the risk of bias associated with individual studies, and assessing the quality of the body of evidence. The importance of managing conflicts of interests and engaging people with expertise in nutrition throughout the systematic review process was also described.

To maximise efficiency, the importance of having clearly defined research questions from the outset was often described. Having clear definitions of the exposures of interest was considered particularly important (e.g., 'intake of foods high in sodium' compared to 'sodium intake' and 'fresh meat' compared to 'processed meat'). For example:

" . . . I remember specific points of discussion around, how is [meat] dealt with? You know, what's red meat? What's white meat? What's processed meat? What's not?" [Participant 17, Review Team].

Some participants described the importance of tailoring the search strategy and the inclusion criteria to the research question so that relevant evidence is not missed or excluded. For example, the inclusion criteria for the date of publication may vary because the evidence that is most suitable for answering a particular research question may have been published during a particular time period. The inclusion criteria for study design may also vary because some study designs are more suitable for addressing particular research questions than others. For example, in relation to evidence on environmental sustainability:

"So we have to be prepared to stand back and look at the way in which evidence is constructed in different disciplines, because what you don't want is to say we're going to have to have a whole lot of systematic reviews [on environmental sustainability], find there's two studies, and then find we can't say anything, because there isn't any research, but there's probably a lot of research, it's just not constructed in that way" [Participant 3, Working Committee].

The importance of selecting risk of bias assessment tools that are appropriate for particular research questions and particular study designs was emphasised. Some participants suggested that risk of bias tools should consider unique factors associated with conducting studies with dietary exposures, including the accurate estimation of dietary intake, and difficulties associated with blinding. For example:

"I think it's always useful to have tools that are appropriate for the study designs and the questions being asked. Most of the risk of bias, certainly the Cochrane Risk of Bias tools, downgrade many nutrition studies, because of the problems with blinding, [but] blinding is quite difficult, so that needs to be managed in some way" [Participant 15, Review Team].

The importance of using the most appropriate method for assessing the quality of the body of evidence was emphasised. There were mixed views on the suitability of the GRADE approach. For example, some participants explained that GRADE allows evidence from observational studies to be upgraded when particular criteria are met, which was described as an advantage. Others argued that alternative approaches may be more suitable for assessing the quality of evidence derived from studies with dietary exposures. For example:

"I think on reflection, it would be good to have a system which is specific to dietary type studies, because you're hardly ever going to get RCTs in this kind of field. They're more likely to be cohort studies or large population studies. And that doesn't make them bad. But when you rate them in traditional systems, they always look like low quality evidence. So I think whatever system that is used going forward, there needs to be a process for rating the evidence provided by those kinds of studies more appropriately" [Participant 20, Review Team].

Some participants described the importance of having a clear process in place to assess the applicability of evidence. For example, it was noted that most dietary patterns evidence is derived from studies conducted in other countries, which may not be applicable to the Australian context:

" . . . my assumption would be that a lot of the [dietary pattern] studies would not be necessarily conducted in Australia, but they may be European or US studies that could be applied to our Australian context. Again, same thing in reverse, there may be some that may not be relevant at all. And we need to have a process in place where we start to work out, how do we deem them as relevant and what would then be appropriate to be considered for the Australian context" [Participant 9, Review Team].

Practicalities associated with the management of actual, potential, and perceived conflicts of interest throughout the systematic review process were discussed. Some participants described the importance of considering conflicts of interest within the evidence base. For example:

"I think something that we've recognised is that the scientific literature has conflicts of interest within it. And we did include sources of funding. When we did the literature review, we extracted data on the sources of funding. But I think that this time, we would pay more careful attention to that, and perhaps highlight that" [Participant 21, Review Team].

The importance of managing conflicts of interest within the systematic review team was also highlighted. For example:

"So I think it's really important in the future that there's a lot of oversight to make sure that we don't have reviewers who have vested interests, [or] funding for other work" [Participant 2, Working Committee].

Some participants described the importance of engaging people with expertise in nutrition throughout the systematic review process on the basis that people without nutrition expertise may misinterpret the evidence. For example:

"I think you do need people who understand nutrition to do it. There's lots of professional, systematic reviewing people, but honestly, if they don't understand nutrition, it can lead to erroneous conclusions" [Participant 21, Review Team].

3.3. Practicalities Associated with Evidence Translation

In relation to the evidence translation stage of dietary guideline development, practicalities associated with evidence integration and stakeholder consultation were described.

3.3.1. Integrating the Evidence

Integrating evidence from multiple systematic reviews with different exposures and outcomes, making decisions about the quality of evidence that is required to inform recommendations, and balancing the potential health, economic, social, and environmental consequences associated with particular recommendations were described as important processes that require professional judgement. For example:

"So theoretically, we might be getting reasonably close to saying, these are the combinations of foods that we think will give the best outcomes in terms of population health. But then you have to ask yourself, if the population is to consume this way, how good are we at providing it? And what is the impact on the environment of us producing it? What is the economic cost? Is this something that we trade in? How important is it from the point of view of the country's GDP? And then the social side of it, which groups in our society actually eat like this? And what are the consequences of us saying they have to eat this other way?" [Participant 3, Working Committee].

Views on whether dietary guideline statements should focus on dietary patterns or on foods were mixed. Some participants supported the development of dietary guideline statements that describe healthy dietary patterns in broad terms (e.g., consuming a diet that is predominantly plant-based but includes small quantities of minimally-processed animal products). Others explained that dietary guideline statements should continue to focus on foods because that is what members of the general public are likely to understand.

Some participants explained that healthy dietary patterns may include small amounts of unhealthy foods, so, by focusing on dietary patterns, there is a risk that messages about unhealthy foods can get lost or used to the advantage of food industry stakeholders. For example:

"I guess one challenge with the dietary patterns approach is that any unhealthy food can be part of a healthy diet pattern, as far as our food industry friends would be concerned. So that's a bit of a downside, in that we play into this rhetoric of, 'it's all about the total diet' when really we do want to be highlighting some foods, all the discretionary foods basically, and having clear messages about those. I think that's where it gets a bit difficult" [Participant 10, Working Committee].

To address some of these challenges, it was suggested that dietary guidelines should continue to focus on foods but also incorporate clear messages about variety, balance, and moderation.

Views on the food classification system that should underpin the dietary guidelines were mixed. Some participants were of the opinion that the current food classification system that includes five core food groups and 'discretionary foods' is not well understood by members of the general public. Other participants highlighted the value of consistent messaging and were concerned that the introduction of a new food classification system could lead to confusion about which foods are 'healthy' or 'unhealthy'. For example:

" . . . I do wonder if as nutrition scientists, we actually should just get to the point where we call out what's not a very healthy food and should be consumed in small amounts. And we all agree on that. If we just keep inventing classification after classification, and spending all our time on that, then it actually may play into the hands of interests that are not so interested in people having a healthy diet" [Participant 21, Review Team].

Some participants supported the development of an Australian-specific version of the NOVA food classification system that classifies foods according to level of processing. Others did not support the use of the NOVA system because they were concerned that consumers may not understand the differences between the four food groups (minimally processed foods, processed culinary ingredients, processed foods, and ultra-processed foods) and, instead, assume that any kind of food processing makes a food 'unhealthy'.

There was limited support for the development of recommendations about eating occasions due to the following risks: statements that promote eating with others might have negative implications for people who live alone; statements that encourage home-prepared meals might have negative implications for people who do not have the capacity to prepare meals at home; concerns that home-prepared meals are not necessarily healthy and that meals eaten away from home are not necessarily unhealthy were expressed; statements about *how* people should eat might take the focus away from statements about *what* people should eat; and statements that focus on eating occasions could be used to the advantage of food industry stakeholders. For example:

" . . . the more that we sort of broaden it out with other messages, the more the food industry can say 'Well, it's all about enjoyment. We have a dietary guideline about enjoying meals with families and family and friends. And here is our McDonald's meal on a Friday' . . . " [Participant 10, Working Committee].

Some participants expressed support for recommendations that focus on health promotion rather than disease prevention and messages that focus on short-term rather than long-term benefits on the basis that these messages may be more appealing to members of the general public. For example:

"I think we all have a bit of a short-term view of things, and we all feel a little bit invincible. So making it closer in terms of timeframes. So feeling well, not getting sick, being able to do the things that you want to do, so being able to enjoy life, particularly as people get older" [Participant 15, Review Team].

The importance of considering the environmental consequences associated with the production and consumption of the foods and dietary patterns that are recommended was emphasised. However, to avoid unintended health consequences, it was also noted that recommendations that encourage lower consumption of meat and dairy (for environmental reasons) should be balanced by recommendations that ensure that nutrient reference values for iron and calcium can be met.

From an equity perspective, the importance of ensuring that the foods and dietary patterns that are recommended are commonly available, accessible, and able to be utilised was emphasised. For example:

> "So there's two layers of evidence, first of all, which foods and which compounds in foods and which combinations of foods are best for health. And there could be a number of different ways you could eat, we know that. But then the second layer is, which of these combinations are achievable in terms of our cuisine, in terms of people's food preferences, in terms of accessibility and sustainability for ensuring equitable food access?" [Participant 16, Review Team].

Some participants described the importance of involving people with expertise in nutrition in the drafting of dietary guideline statements and the development of food guides. For example:

> " ... the food guide is essential in taking all of those technical reviews and all of the actual guidelines and translating that for the public. So if people with all of that skill and expertise are not involved in that really important next step, it sort of negates the process. Because if you can't get the message across properly, then there's no point in having dietary guidelines..." [Participant 16, Review Team].

3.3.2. Consulting Stakeholders

To inform the wording of dietary guideline statements and the development of food guides and other resources, the need for a comprehensive and well-resourced consultation process was often described. For example:

> " ... we always think that we know best, and we actually don't know best, we don't know what the consumer and what the person out there on the street, what message they receive. And we need to just be investing much more time in trying to understand that. But we're not, we're going to go through the whole process of doing a million [systematic reviews], synthesising the evidence, which is all good and well, but then we fail at the most important step, which is understanding how consumers understand" [Participant 16, Review Team].

The importance of engaging people from a diverse range of socioeconomic and cultural backgrounds in the consultation process and managing the influence of food industry stakeholders with conflicts of interest was often described. Some participants explained that a robust evidence review process can reduce the influence that submissions from stakeholders with conflicts of interest have on the final version of the dietary guidelines. For example:

> "When you can say, look, we're basing it on proper systematic reviews. And that was the good thing about the last one [...] we could say, well, this was the evidence that we were using and this was why ... " [Participant 1, Review Team].

4. Discussion

The aim of this study was to explore perceived advantages, disadvantages, and practicalities associated with the synthesis and translation of evidence from nutrient-based, food-based, and dietary patterns research in dietary guideline development. To inform the next iteration of the ADGs, there was support for reviewing evidence on a range of dietary exposures (including dietary patterns, foods and food groups, nutrients and food components, and eating occasions) and health outcomes. In response to increasing environmental and economic threats to food security in Australia, the importance of reviewing the latest evidence on environmental sustainability and equity was described. At the evidence synthesis stage, practicalities associated with planning the evidence review and conducting original systematic reviews were discussed. At the evidence translation stage, practicalities associated with integrating the evidence and consulting stakeholders were described.

The expert committee is typically responsible for identifying and prioritizing the research questions that will be addressed in the evidence review [1,2]. In this study, participants consistently described dietary patterns as important exposures, whereas views on the advantages of reviewing evidence on eating occasions were mixed. Reviews of evidence on dietary patterns and eating occasions have been conducted to inform dietary guidelines in other countries. For example, a series of systematic reviews on dietary patterns and body weight, cardiovascular disease, and type 2 diabetes was conducted to inform the 2015 Dietary Guidelines for Americans [53]. To inform the 2020 Dietary Guidelines for Americans, evidence on the associations between eating frequency and health outcomes and eating frequency and diet quality was also reviewed [11]. The current Dietary Guidelines for the Brazilian Population were informed by a review of evidence on modes of eating (such as eating at regular times and eating with other people) as determinants of healthy diets [54]. A combination of existing and original systematic reviews of relevant evidence on dietary patterns and eating occasions (as exposures) and a range of health outcomes could be used to inform future ADGs.

Research questions about dietary exposures and long-term health outcomes (e.g., chronic disease incidence) can be difficult to address using RCTs due to challenges associated with compliance to dietary interventions and the costs associated with delivering interventions over long periods of time [7,16,55]. Prospective cohort studies are more suitable for addressing these types of research questions but are subject to risk of bias associated with confounding [14,55]. To ensure that the quality of relevant evidence from observational studies is assessed accurately, study participants described the importance of using appropriate risk of bias tools, including tools that consider unique challenges associated with conducting observational studies with dietary exposures (e.g., accurate estimation of dietary intake and difficulties associated with blinding). In Australia, the latest NHMRC Guidelines for Guidelines handbook highlights the importance of using an appropriate risk of bias assessment tool, noting that, "depending on the type of research question, strong observational studies can at times provide more reliable evidence than flawed randomised trials" [56]. Cochrane risk of bias tools exist for randomised trials (ROB2) [57] and non-randomised studies of interventions (ROBINS-I) [58]. These tools consider the sources of bias associated with particular study designs, but were not developed specifically for assessing studies with dietary exposures. In contrast, the Risk of Bias for Nutrition Observational Studies (RoB-NObs) Tool was created by the USDA's Nutrition Evidence Systematic Review team and used in dietary guideline development [11]. In the development of future ADGs, the use of tools that have been developed to assess the risk of bias associated with observational studies focused on dietary exposures should be considered.

The importance of using an appropriate tool to assess the quality of evidence from observational studies with dietary exposures was described by study participants. Using the GRADE approach, evidence from RCTs is rated more highly than evidence from non-randomised studies from the outset, regardless of the research question [8,23]. The NHRMC guidelines recommend the use of GRADE for assessing the certainty of the evidence but acknowledge the ongoing debate about whether GRADE is the most appropriate system to use in public health guideline development [59]. This debate is also evident in nutrition science literature. For example, Zeraatkar et al. recommend the use of GRADE in dietary guideline development on the basis that this system is the most rigorous [6]. Conversely, Tobias et al. acknowledge the strengths of GRADE but argue that alternative approaches that may be more appropriate for assessing the certainty of evidence from observational studies with dietary exposures should also be considered [22]. In the development of future ADGs, the use of alternative evidence assessment approaches such as the Hierarchies of Evidence Applied to Lifestyle Medicine (HEALM) approach could be considered for research questions that cannot be addressed using RCTs [24].

At the evidence translation stage, integrating the evidence, making decisions about the quality of evidence that is required to inform recommendations, and balancing the potential consequences of recommendations were described as processes that require professional judgement. Blake et al. analysed dietary guideline development processes in 32 countries [5]. They found that in 28 countries, the dietary guidelines committee used an unstructured consensus process to translate the evidence into recommendations. In the remaining four countries, information on the evidence translation process was not available [5]. The need for increased transparency at the evidence translation stage of dietary guideline development has been recognised [60,61]. A number of frameworks now exist to guide decision making in the translation of evidence into recommendations, including the GRADE Evidence to Decision Framework and the WHO-INTEGRATE framework [62,63]. Each of these frameworks includes criteria on the quality of evidence and balancing health benefits and harms. The WHO-INTEGRATE framework also incorporates social, economic, and environmental dimensions [63]. In the development of future ADGs, the use of a suitable framework to guide the translation of evidence into recommendations should be considered.

Figure 2 provides an overview of the practicalities associated with evidence synthesis and translation in dietary guideline development, summarising the results of this study in the context of approaches that have been described as best practice in the literature.

By exploring the ideas and opinions of people with first-hand experience, this study provides insights into the dietary guideline development process in Australia. Although the sample size was not based on the concept of information power, it is likely that a high level of information power was reached because interviews were conducted with a purposive sample of participants, including 6 out of the 12 eligible members of the Working Committee and 16 out of the 36 eligible members of the Review Team, and the quality of the dialogue was strong [45]. However, it is possible that certain information was not captured because not everyone who was eligible to participate was interviewed. To increase rigor in the qualitative approach, strategies to enhance reflexivity were used, and a subset of transcripts were analysed by multiple researchers [39,42]. A limitation of this study is that dietary guidelines are designed to be context-specific, which limits the transferability of results [42]. However, the practicalities associated with evidence synthesis and translation that were identified may be relevant in other contexts.

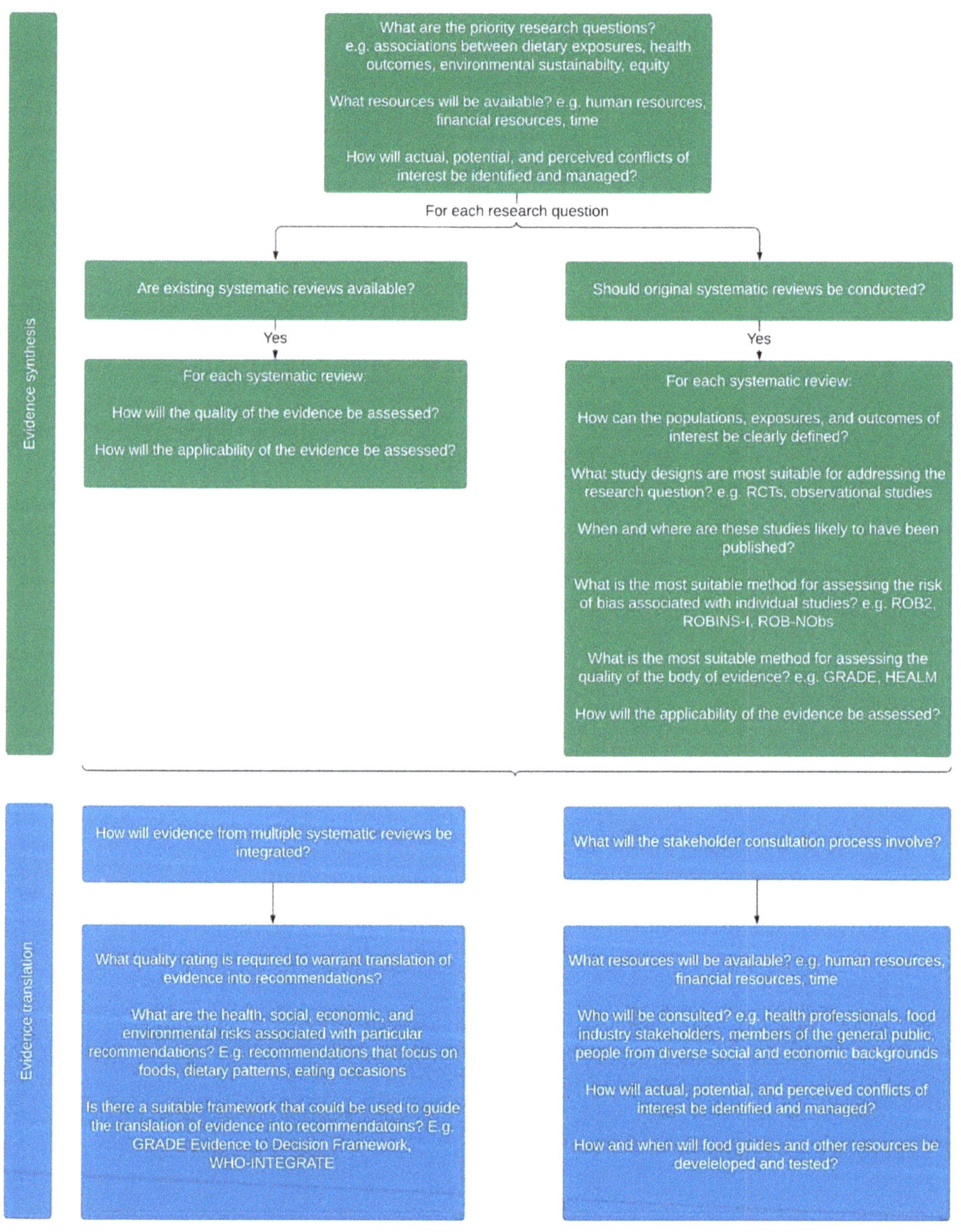

Figure 2. An overview of the practicalities associated with evidence synthesis and translation in dietary guideline development.

5. Conclusions

Dietary guidelines are important nutrition policy reference standards that should be informed by the best available evidence. This study explored perceived advantages, disadvantages, and practicalities associated with the synthesis and translation of evidence in dietary guideline development. To inform future ADGs, there was support for reviewing evidence on the associations between a range of dietary exposures (including dietary patterns, foods and food groups, nutrients and food components, and eating occasions) and health outcomes along with the latest Australian evidence on environmental sustainability and equity. To ensure that the best available evidence is translated into dietary guidelines, the most appropriate evidence review methods should be selected based on the exposures and outcomes of interest.

Supplementary Materials: The following are available online at https://www.mdpi.com/article/10.3390/nu13113748/s1, Table S1: semi-structured interview guide.

Author Contributions: Conceptualization, K.W., S.A.M. and M.A.L.; methodology, K.W., S.A.M. and M.A.L.; formal analysis, K.W. and C.R.; writing—original draft preparation, K.W.; writing—review and editing, K.W., S.A.M., M.A.L. and C.R.; supervision, S.A.M. and M.A.L.; project administration, K.W.; funding acquisition, K.W. and S.A.M. All authors have read and agreed to the published version of the manuscript.

Funding: This study was funded by an Australian Government Research Training Program Scholarship (K.W.).

Institutional Review Board Statement: This study was conducted according to the guidelines of the Declaration of Helsinki and approved by the Deakin University Human Ethics Advisory Group (HEAG-H 35_2020 approved 4 May 2020).

Informed Consent Statement: Informed consent was obtained from all subjects involved in the study.

Data Availability Statement: The data are not available due to privacy restrictions.

Conflicts of Interest: K.W. was a member of the Dietitians Australia (DA) Australian Dietary Guidelines Review Working Group (from March 2021) and the DA Food and Environment Interest Group Leadership Team (from August 2020). S.A.M. was appointed Chair of the NHMRC Dietary Guidelines Expert Committee from September 2021. M.A.L. was a member of the NHMRC Working Committee for the review of the Australian Dietary Guidelines (2008–2013). The funders had no role in the design of the study; in the collection, analyses, or interpretation of data; in the writing of the manuscript; or in the decision to publish the results.

References

1. FAO; WHO. *Preparation and Use of Food-Based Dietary Guidelines. Report of a Joint FAO/WHO Consultation*; World Health Organization: Geneva, Switzerland, 1998; Available online: http://www.who.int/iris/handle/10665/42051 (accessed on 7 June 2021).
2. FAO. *Developing Food Based Dietary Guidelines. A Manual from the English-Speaking Caribbean*; Food and Agriculture Organization of the United Nations: Rome, Italy, 2007; Available online: http://www.fao.org/3/ai800e/ai800e00.htm (accessed on 7 June 2021).
3. FAO. Food-Based Dietary Guidelines. Available online: http://www.fao.org/nutrition/education/food-dietary-guidelines/home/en/ (accessed on 7 April 2021).
4. Wijesinha-Bettoni, R.; Khosravi, A.; Ramos, A.I.; Sherman, J.; Hernandez-Garbanzo, Y.; Molina, V.; Vargas, M.; Hachem, F. A snapshot of food-based dietary guidelines implementation in selected countries. *Glob. Food Sec.* **2021**, *29*. [CrossRef]
5. Blake, P.; Durao, S.; Naude, C.E.; Bero, L. An analysis of methods used to synthesize evidence and grade recommendations in food-based dietary guidelines. *Nutr. Rev.* **2018**, *76*, 290–300. [CrossRef] [PubMed]
6. Zeraatkar, D.; Johnston, B.C.; Guyatt, G. Evidence collection and evaluation for the development of dietary guidelines and public policy on nutrition. *Annu. Rev. Nutr.* **2019**, *39*, 227–247. [CrossRef]
7. Tapsell, L.C.; Neale, E.P.; Satija, A.; Hu, F.B. Foods, nutrients, and dietary patterns: Interconnections and implications for dietary guidelines. *Adv. Nutr.* **2016**, *7*, 445–454. [CrossRef]
8. Bero, L.A.; Norris, S.L.; Lawrence, M.A. Making nutrition guidelines fit for purpose. *BMJ* **2019**, *365*, 1579. [CrossRef]
9. Mozaffarian, D.; Forouhi, N.G. Dietary guidelines and health—Is nutrition science up to the task? *BMJ* **2018**, *360*, 822. [CrossRef] [PubMed]
10. Mozaffarian, D.; Rosenberg, I.; Uauy, R. History of modern nutrition science—Implications for current research, dietary guidelines, and food policy. *BMJ* **2018**, *361*, 1–6. [CrossRef] [PubMed]

11. Dietary Guidelines Advisory Committee. *Scientific Report of the 2020 Dietary Guidelines Advisory Committee: Advisory Report to the Secretary of Agriculture and the Secretary of Health and Human Services*; United States Department of Agriculture: Washington, DC, USA, 2020. Available online: https://www.dietaryguidelines.gov/2020-advisory-committee-report (accessed on 14 June 2021).

12. Kumanyika, S.; Afshin, A.; Arimond, M.; Lawrence, M.; McNaughton, S.A.; Nishida, C. Approaches to Defining Healthy Diets: A Background Paper for the International Expert Consultation on Sustainable Healthy Diets. *Food Nutr. Bull.* **2020**, *41*, 7S–30S. [CrossRef]

13. McNaughton, S.A. *Dietary Patterns*; International Life Sciences Institute (ILSI): Washington, DC, USA, 2020; Volume 2, pp. 235–248.

14. Schulze, M.B.; Martínez-González, M.A.; Fung, T.T.; Lichtenstein, A.H.; Forouhi, N.G. Food based dietary patterns and chronic disease prevention. *BMJ* **2018**, *361*, 2396. [CrossRef]

15. WCRF; AICR. *Continuous Update Project Expert Report 2018. Judging the Evidence*; World Cancer Research Fund and American Institute for Cancer Research: London, UK, 2018; Available online: https://www.wcrf.org/wp-content/uploads/2021/02/judging-the-evidence.pdf (accessed on 14 June 2021).

16. Jacobs, D.R.; Tapsell, L.C.; Temple, N.J. Food synergy: The key to balancing the nutrition research effort. *Public Health Rev.* **2012**, *33*, 507–529. [CrossRef]

17. Reedy, J.; Subar, A.F.; George, S.M.; Krebs-Smith, S.M. Extending methods in dietary patterns research. *Nutrients* **2018**, *10*, 571. [CrossRef]

18. Jacobs, D.R.; Tapsell, L.C. Food synergy: The key to a healthy diet. *Proc. Nutr. Soc.* **2013**, *72*, 200–206. [CrossRef]

19. Cespedes, E.M.; Hu, F.B. Dietary patterns: From nutritional epidemiologic analysis to national guidelines. *Am. J. Clin. Nutr.* **2015**, *101*, 899–900. [CrossRef]

20. WHO. *Handbook for Guideline Development*; World Health Organization: Geneva, Switzerland, 2014; Available online: http://apps.who.int/medicinedocs/documents/s22083en/s22083en.pdf (accessed on 20 August 2021).

21. Higgins, J.P.T.; Thomas, J.; Chandler, J.; Cumpston, M.; Li, T.; Page, M.J.; Welch, V.A. Cochrane Handbook for Systematic Reviews of Interventions Version 6.2 (Updated February 2021); Cochrane: 2021. Available online: www.training.cochrane.org/handbook (accessed on 11 October 2021).

22. Tobias, D.K.; Wittenbecher, C.; Hu, F.B. Grading nutrition evidence: Where to go from here? *Am. J. Clin. Nutr.* **2021**. [CrossRef] [PubMed]

23. Schunemann, H.J.; Cuello, C.; Akl, E.A.; Mustafa, R.A.; Meerpohl, J.J.; Thayer, K.; Morgan, R.L.; Gartlehner, G.; Kunz, R.; Katikireddi, S.V.; et al. GRADE guidelines: 18. How ROBINS-I and other tools to assess risk of bias in nonrandomized studies should be used to rate the certainty of a body of evidence. *J. Clin. Epidemiol.* **2018**. [CrossRef] [PubMed]

24. Katz, D.L.; Karlsen, M.C.; Chung, M.; Shams-White, M.M.; Green, L.W.; Fielding, J.; Saito, A.; Willett, W. Hierarchies of evidence applied to lifestyle Medicine (HEALM): Introduction of a strength-of-evidence approach based on a methodological systematic review. *BMC Med. Res. Methodol.* **2019**, *19*, 178. [CrossRef]

25. Truswell, A.S. Dietary guidelines: Theory and practice. *Proc. Nutr. Soc. Aust.* **1995**, *19*, 1–10.

26. NHMRC. *Dietary Guidelines for Australians*; National Health and Medical Research Council: Canberra, Australia, 1992.

27. NHMRC. *Dietary Guidelines for Children and Adolescents*; National Health and Medical Research Council: Canberra, Australia, 1995.

28. NHMRC. *Dietary Guidelines for Older Australians*; National Health and Medical Research Council: Canberra, Australia, 1999.

29. NHMRC. *Dietary Guidelines for Children and Adolescents in Australia*; National Health and Medical Research Council: Canberra, Australia, 2003.

30. NHMRC. *Dietary Guidelines for Australian Adults*; National Health and Medical Research Council: Canberra, Australia, 2003.

31. NHMRC. *Australian Dietary Guidelines*; National Health and Medical Research Council: Canberra, Australia, 2013. Available online: https://www.nhmrc.gov.au/guidelines/publications/n55 (accessed on 10 June 2021).

32. NHMRC. *A Modelling System to Inform the Revision of the Australian Guide to Healthy Eating*; Commonwealth of Australia: Canberra, Australia, 2011. Available online: https://www.eatforhealth.gov.au/sites/default/files/files/public_consultation/n55a_dietary_guidelines_food_modelling_111216.pdf (accessed on 10 June 2021).

33. Wingrove, K.; Lawrence, M.A.; McNaughton, S.A. Dietary patterns, foods and nutrients: A descriptive analysis of the systematic reviews conducted to inform the Australian Dietary Guidelines. *Nutr. Res. Rev.* **2020**, 1–8. [CrossRef]

34. NHMRC. *Review of the Australian Dietary Guidelines: Living Well for Longer*; National Health and Medical Research Council: Canberra, Australia, 2020. Available online: https://www.health.gov.au/ministers/the-hon-greg-hunt-mp/media/review-of-the-australian-dietary-guidelines-living-well-for-longer (accessed on 10 June 2021).

35. Mann, J.I. Evidence-based nutrition: Does it differ from evidence-based medicine? *Ann. Med.* **2010**, *42*, 475–486. [CrossRef] [PubMed]

36. Allman-Farinelli, M.; Byron, A.; Collins, C.; Gifford, J.; Williams, P. Challenges and lessons from systematic literature reviews for the Australian dietary guidelines. *Aust. J. Prim. Health* **2014**, *20*, 236–240. [CrossRef]

37. Sandelowski, M. Whatever happened to qualitative description? *Res. Nurse Health* **2000**, *23*, 334–340. [CrossRef]

38. Sandelowski, M. What's in a name? Qualitative description revisited. *Res. Nurs. Health* **2010**, *33*, 77–84. [CrossRef]

39. Hansen, E.C. *Successful Qualitative Health Research: A Practical Introduction*; Allen & Unwin: Crows Nest, Australia, 2006.

40. Swift, J.A.; Tischler, V. Qualitative research in nutrition and dietetics: Getting started. *J. Hum. Nutr. Diet.* **2010**, *23*, 559–566. [CrossRef]

41. Tong, A.; Sainsbury, P.; Craig, J. Consolidated criteria for reporting qualitative research (COREQ): A 32-item checklist for interviews and focus groups. *Int. J. Qual. Health Care* **2007**, *16*, 349–357. [CrossRef]
42. Draper, A.; Swift, J.A. Qualitative research in nutrition and dietetics: Data collection issues. *J. Hum. Nutr. Diet.* **2011**, *24*, 3–12. [CrossRef] [PubMed]
43. FAO. Food-based dietary guidelines—Australia. Available online: http://www.fao.org/nutrition/education/food-based-dietary-guidelines/regions/countries/australia/es/ (accessed on 11 October 2021).
44. NHMRC. *A Review of the Evidence to Address Targeted Questions to Inform the Revision of the Australian Dietary Guidelines*; National Health and Medical Research Council: Canberra, Australia, 2011. Available online: https://www.eatforhealth.gov.au/sites/default/files/files/the_guidelines/n55d_dietary_guidelines_evidence_report.pdf (accessed on 20 August 2021).
45. Malterud, K.; Siersma, V.D.; Guassora, A.D. Sample size in qualitative interview studies: Guided by information power. *Qual. Health Res.* **2016**, *26*, 1753–1760. [CrossRef] [PubMed]
46. Braun, V.; Clarke, V. To saturate or not to saturate? Questioning data saturation as a useful concept for thematic analysis and sample-size rationales. *Qual. Res. Sport Exerc. Health* **2019**, 1–16. [CrossRef]
47. Jacob, S.A.; Furgerson, S.P. Writing interview protocols and conducting interviews: Tips for students new to the field of qualitative research. *Qual. Rep.* **2012**, *17*, 1–10.
48. Orb, A.; Eisenhauer, L.; Wynaden, D. Ethics in qualitative research. *J Nurs. Scholarsh.* **2001**, *33*, 93–96. [CrossRef]
49. Burnard, P.; Gill, P.; Stewart, K.; Treasure, E.; Chadwick, B. Analysing and presenting qualitative data. *Br. Dent. J.* **2008**, *204*, 429–432. [CrossRef]
50. Green, J.; Willis, K.; Hughes, E.; Small, R.; Welch, N.; Gibbs, L.; Daly, J. Generating best evidence from qualitative research: The role of data analysis. *Aust, N Z J. Public Health* **2007**, *31*, 545–550. [CrossRef]
51. QSR International Pty Ltd. NVivo (Version 12 Plus). 2018. Available online: https://www.qsrinternational.com/nvivo-qualitative-data-analysis-software/home/ (accessed on 10 October 2021).
52. Stok, F.M.; Renner, B.; Allan, J.; Boeing, H.; Ensenauer, R.; Issanchou, S.; Kiesswetter, E.; Lien, N.; Mazzocchi, M.; Monsivais, P.; et al. Dietary behavior: An interdisciplinary conceptual analysis and taxonomy. *Front. Psychol.* **2018**, *9*, 1689. [CrossRef] [PubMed]
53. USDA. *A Series of Systematic Reviews on the Relationship Between Dietary Patterns and Health Outcomes*; United States Department of Agriculture: Washington, DC, USA, 2014. Available online: https://nesr.usda.gov/sites/default/files/2019-06/DietaryPatternsReport-FullFinal2.pdf (accessed on 20 August 2021).
54. Louzada, M.L.d.C.; Canella, D.S.; Jaime, P.C.; Monteiro, C.A. *Food and Health: The Scientific Evidence Informing the Dietary Guidelines for the Brazilian Population*; Faculdade de Saúde Pública and Universidade de São Paulo: Sao Paulo, Brazil, 2019; Available online: http://www.livrosabertos.sibi.usp.br/portaldelivrosUSP/catalog/view/402/355/1436-1 (accessed on 20 August 2021).
55. Satija, A.; Yu, E.; Willett, W.C.; Hu, F.B. Understanding nutritional epidemiology and its role in policy. *Adv. Nutr.* **2015**, *6*, 5–18. [CrossRef] [PubMed]
56. NHMRC. *Guidelines for Guidelines: Assessing Risk of Bias*; National Health and Medical Research Council: Canberra, Australia, 2019. Available online: https://www.nhmrc.gov.au/guidelinesforguidelines/develop/assessing-risk-bias (accessed on 1 September 2021).
57. Sterne, J.A.C.; Savovic, J.; Page, M.J.; Elbers, R.G.; Blencowe, N.S.; Boutron, I.; Cates, C.J.; Cheng, H.Y.; Corbett, M.S.; Eldridge, S.M.; et al. RoB 2: A revised tool for assessing risk of bias in randomised trials. *BMJ* **2019**, *366*, l4898. [CrossRef]
58. Sterne, J.A.; Hernan, M.A.; Reeves, B.C.; Savovic, J.; Berkman, N.D.; Viswanathan, M.; Henry, D.; Altman, D.G.; Ansari, M.T.; Boutron, I.; et al. ROBINS-I: A tool for assessing risk of bias in non-randomised studies of interventions. *BMJ* **2016**, *355*, i4919. [CrossRef]
59. NHMRC. *Guidelines for Guidelines: Assessing Certainty of Evidence*; National Health and Medical Research Council: Canberra, Australia, 2019. Available online: https://www.nhmrc.gov.au/guidelinesforguidelines/develop/assessing-certainty-evidence (accessed on 1 September 2021).
60. Bero, L. Developing reliable dietary guidelines. *BMJ* **2017**, *359*, 4845. [CrossRef] [PubMed]
61. Nestle, M. Perspective: Challenges and controversial issues in the Dietary Guidelines for Americans, 1980–2015. *Adv Nutr.* **2018**, *9*, 148–150. [CrossRef] [PubMed]
62. Alonso-Coello, P.; Schunemann, H.J.; Moberg, J.; Brignardello-Petersen, R.; Akl, E.A.; Davoli, M.; Treweek, S.; Mustafa, R.A.; Rada, G.; Rosenbaum, S.; et al. GRADE Evidence to Decision (EtD) frameworks: A systematic and transparent approach to making well informed healthcare choices. 1: Introduction. *BMJ* **2016**, *353*, 2016. [CrossRef] [PubMed]
63. Rehfuess, E.A.; Stratil, J.M.; Scheel, I.B.; Portela, A.; Norris, S.L.; Baltussen, R. The WHO-INTEGRATE evidence to decision framework version 1.0: Integrating WHO norms and values and a complexity perspective. *BMJ Glob. Health* **2019**, *4*, e000844. [CrossRef]

 *nutrients*

Article

Number of Days Required to Estimate Habitual Vegetable Variety: A Cross-Sectional Analysis Using Dietary Records for 7 Consecutive Days

Ryoko Kurisaki and Osamu Kushida *

Department of Nutrition and Life Sciences, School of Food and Nutritional Sciences, University of Shizuoka, 52-1 Yada, Suruga-ku, Shizuoka 422-8526, Japan; f17111@u-shizuoka-ken.ac.jp
* Correspondence: kushida@u-shizuoka-ken.ac.jp; Tel.: +81-54-264-5832

Abstract: The aim of this cross-sectional study was to examine the number of days required to estimate habitual vegetable variety by conducting a multiday, dietary record. Sixty respondents from three groups in Japan (rural residents, general students, and nutrition students) participated in the study using a self-administered questionnaire in September 2018. To measure vegetable variety, the number of different vegetables consumed was extracted from the dietary records of seven consecutive days. Differences in the number of vegetables consumed and the capture proportion over seven consecutive days between groups were examined using repeated measures analysis of variance and one-way analysis of variance. The vegetable variety between each day was also compared using Pearson's correlation coefficient. The vegetable variety based on dietary records for seven consecutive days confirmed the differences between groups by repeated measurements ($p = 0.013$). However, there was no significant difference among groups in the capture proportion per survey day based on seven consecutive days. Furthermore, there were significant correlations between the number of vegetables consumed over seven consecutive days and that consumed on two or more days ($r > 0.50$, $p < 0.01$) and especially three or more days in all groups ($r > 0.70$, $p < 0.001$). The present study suggested that a dietary survey over two or more days could provide an estimate of habitual vegetable variety.

Keywords: vegetable; variety; habitual; dietary records; capture proportion

Citation: Kurisaki, R.; Kushida, O. Number of Days Required to Estimate Habitual Vegetable Variety: A Cross-Sectional Analysis Using Dietary Records for 7 Consecutive Days. *Nutrients* **2022**, *14*, 56. https://doi.org/10.3390/nu14010056

Academic Editors: Colin Bell and Penny Love

Received: 10 November 2021
Accepted: 21 December 2021
Published: 24 December 2021

Publisher's Note: MDPI stays neutral with regard to jurisdictional claims in published maps and institutional affiliations.

1. Introduction

The association of the amount of fruits and vegetables (FV) consumed with chronic diseases such as cardiovascular disease and cancer has been analyzed by many nutritional epidemiological studies [1–3]. Because inadequate intake of FV contributes to many chronic diseases and global excess mortality, the World Health Organization (WHO) recommends increasing the amount of FV consumed globally [4].

Recently, many studies have examined the association between health status and FV variety rather than consumed amount of FV. For example, vegetable variety has been reported to be positively correlated with overall diet quality [5] and higher cognitive function [6]. Vegetable variety has also been reported to reduce the onset risk of chronic diseases such as type 2 diabetes [7] and lung cancer [8]. Regarding cognitive function [6] and lung cancer [8], because vegetable variety has been more strongly associated with health status than the consumed amount of vegetables, it is important to increase vegetable variety. The American and Australian dietary guidelines also make recommendations to increase the vegetable variety in subgroups, not just the consumed amount [9,10].

According to a scoping review summarizing the operationalization of FV variety [11], many studies have used a food frequency questionnaire (FFQ) or 24-h dietary recall (24-h recall) to measure FV variety. In general, an FFQ is intended to comprehensively assess the amount of nutrients and foods consumed [12] and is less likely to be developed with the intention of measuring FV variety. In addition, because most of the time frames in studies

---

that have measured FV variety using 24-h recall are 1–2 days, habitual understanding is limited.

In Japan, the strategic importance of assessing vegetable variety is higher than that of assessing fruit variety because more than 50% of those aged 20–49 years were found to have an intake of 0 g/day of fruit [13]. Although some studies have examined the characteristics of the increasing trend in the number of different foods consumed according to multiday dietary records [14,15], there is limited information about the number of different vegetables consumed. By estimating the capture proportion for each survey day and the strength of the correlation between survey days for the number of vegetables consumed, the validity of each survey day can be examined. Therefore, the aim of this study was to examine the number of survey days required to estimate habitual vegetable variety by conducting a multiday dietary record.

2. Materials and Methods

2.1. Participants

A cross-sectional study using an anonymous self-administered questionnaire was conducted in September 2018. The setting for this study was the following three communities in Japan: residents of Village A in a mountainous region of Niigata Prefecture (rural residents), students belonging to a general university in City B in Niigata Prefecture (general students), and fourth-year students belonging to a university for training registered dietitians in Town C in Nara Prefecture (nutrition students). According to the Census of Agriculture and Forestry [16], the forest area per total land is 62.0% in Village A, 7.5% in City B, and 1.4% in Town C. The total area of cultivated land under management per non-forest area is 32.3% in Village A, 44.0% in City B, and 18.5% in Town C.

The eligibility criteria for participants were adults aged 20 years or older who resided within the prefecture of the setting area. The population of participants consisted of approximately 400 rural residents in Village A, 4000 general students in City B, and 40 nutrition students in Town C. Rural residents were recruited through the residents' association, general students were recruited through the students who were doing fieldwork in Village A, and nutrition students were directly recruited by the research staff. The sample size was set at approximately 20 participants for each community because in a previous study that evaluated food variety, each group had 24 participants [14]; in addition, the number of participants was limited to 20 after consultation with representatives of rural residents. A total of 67 questionnaires were distributed directly to participants, and 63 were collected by mail (response rate: 94.0%).

When the questionnaires were distributed, a request for the participants' cooperation was enclosed; this request described the purpose and methods of this study and ethical considerations. The participants provided their informed consent by submitting the questionnaire. This study was conducted after review and approval by the Research Ethics Committee of Kio University of the last author's previous institution (Approval No. H30-31).

2.2. Measures

Qualitative dietary records were maintained for seven consecutive days. Although a larger number of survey days is better for habitual understanding, more survey days leads to a lower rate of participant cooperation. Because it has been suggested that one-week dietary records (seven consecutive days) are best treated statistically as a single measure [17], the number of consecutive days was set at seven in this study. The recorded items were meal start time, name of the dish, name of the food, and information about the commercial products (product name, seller name, and store name). Regarding the time of recording, responses were recorded during the meal or after the meal (with participants either taking photos or forgetting to take photos). Although it is better to record during the meal, if participants completed the dietary record after the meal, they were asked to take photos for their own confirmation, which helped them to reduce food omissions. Since the purpose of this study was to investigate vegetable variety, water and tea were not recorded to reduce participant burden.

In principle, vegetable items were extracted from the items recorded by the participants on the dietary record form. If only the name of the dish was recorded (e.g., curry (Japanese curry and rice) or gyōza (pan-fried dumplings)), foods with a high frequency of use in high-rank web search recipes were adopted. If there was a record of the product or restaurant name that was purchased or consumed along with the dish name, the actual foods were identified by web search. For vegetable juice drinks, the vegetable with the highest content was extracted when the vegetable juice contained 10% or more. Vegetable condiments (e.g., tomato ketchup and ginger paste) were classified in the food group "seasonings and spices," which includes foods with the same name [18], and were not counted as vegetable items. Vegetable seasonings in cup noodles were classified in the food group "cereals," which includes instant foods with included condiments [18], and were also not counted as vegetable items. The vegetable items were extracted and confirmed by two students belonging to a university for training registered dietitians and then reconfirmed by one registered dietitian to increase the validity.

Vegetable variety was defined as the number of different vegetables consumed extracted by the above method. Vegetables were extracted by referring to the Standards Tables of Food Composition in Japan-2015-(Seventh Revised Edition) [18]. For the extraction, vegetables with the same classification but with different food names were defined as different items, while the same vegetables with different cooking and processing methods were defined as the same item. For example, in the case of vegetables classified as peas, pea sprouts and snow peas were counted as different items, while boiled and frozen green peas were counted as the same item (Table 1). Vegetables with different food names but in the same category and differing only in processing method (e.g., "daikon," "kiriboshi-daikon," and "pickles") were counted as the same item.

Table 1. Example of extraction of the number of vegetables consumed.

Item No. [1]	Food and Description [1]	Ex. 1	Ex. 2	Ex. 3	Ex. 4
	(Peas)				
	Pea sprouts				
06019	Stem and leaves, raw				
06329	Sprouts, raw				
06330	Sprouts, boiled	v			
06331	Sprouts, sautéed				
	Snow peas				
06020	Immature pods, raw				
06021	Immature pods, boiled	v			
	Snap peas				
06022	Immature pods, raw				
	Green peas				
06023	Raw				
06024	Boiled		v		
06025	Frozen		v		
06026	Canned in brine				
	(Japanese radish, Daikon)				
	Daikon, sprouts				
06128	Sprouts, raw			v	
	Daikon, cultivar for leaf use				
06129	Leaves, raw				
	Daikon				
06130	Leaves, raw				
06131	Leaves, boiled			v	
06132	Root with skin, raw				v

Table 1. *Cont.*

Item No. [1]	Food and Description [1]	Ex. 1	Ex. 2	Ex. 3	Ex. 4
06133	Root with skin, boiled				
06134	Root without skin, raw				
06135	Root without skin, boiled				
	Kiriboshi-daikon [2]				
06136	Raw				
06334	Rehydrated and boiled				v
06335	Rehydrated and sautéed				
	Pickles				
06137	*Nukamiso-zuke* [3]				v
06138	*Takuan-zuke* [4]				
	Total Items	2	1	2	1

[1] Standards Tables of Food Composition in Japan-2015-(Seventh Revised Edition) [18], [2] Cut and dried daikon root, [3] Pickled in salty rice bran paste, [4] Pickled with rice bran and salt, made of salted daikon.

The following sociodemographic characteristics of the participants were collected: age, gender, height, weight, and number of people living in the same residence. Height (m) and weight (kg) were self-reported and used to calculate body mass index (kg/m^2). In addition, the amount of vegetable consumption has been found to be strongly correlated with vegetable variety in several studies [19–21]. Therefore, the questionnaire also asked for the habitual consumption of the number of vegetable servings [22], which was correlated with the amount of vegetable consumption assessed by the diet history questionnaire. The number of vegetable servings per day was answered on a 5-point ordinal scale from "Very few" to "≥7 servings." The size of a vegetable serving was defined as approximately 70 g salad or boiled greens and was explained using photo images.

2.3. Data Analysis

Differences in sociodemographic characteristics, and the number of vegetable servings per day between groups were examined using one-way analysis of variance for continuous data (age and body mass index), the chi-square test for categorical data (gender and living alone), and the Kruskal-Wallis test for ordinal data (number of vegetable servings). The number of vegetables consumed was determined to be normally distributed by histogram and a normal probability plot. Therefore, differences in vegetable variety on seven consecutive days between groups were examined using repeated measures analysis of variance. The capture proportion of the number of vegetables consumed per survey day based on seven consecutive days was compared using one-way analysis of variance. A Bonferroni correction for multiple comparisons was applied for these individual comparisons. Vegetable variety by the number of survey days was compared by Pearson's correlation coefficient.

Furthermore, the maximum theoretical number of vegetables consumed was calculated based on the method of Asato et al. [15], which examined the number increase in consumed food items. First, the authors took the reciprocal of the number of survey days on the horizontal axis and the reciprocal of the mean number of vegetables consumed in each group on the vertical axis and created an approximately straight line from each plot. Then, the maximum theoretical number was determined by the reciprocal of the intercept on the vertical axis. Although the estimation of such double reciprocal plots is considered unreliable [23], since the maximum theoretical number can be estimated, they were presented as reference values.

IBM SPSS Statistics 27 (IBM Japan, Ltd., Tokyo, Japan) was used for all statistical analyses. The level of significance was set at $p < 0.05$ (two-sided test).

3. Results

Of the 63 respondents, three were excluded because they did not provide sociodemographic information or the names of foods. Thus, a total of 60 participants (16 rural residents, 17 general students, and 27 nutrition students) who were assessed for all seven days were included in the analysis. The sociodemographic characteristics of the participants are shown in Table 2. In terms of the groups with the highest means and proportions for the sociodemographic characteristics, rural residents had the highest mean age, at 59.4 years old; nutrition students had the highest proportion of women, at 81.5%; rural residents had the highest mean body mass index, at 23.1 kg/m^2; and general students had the highest proportion of those living alone, at 76.5% (all $p < 0.05$). The number of vegetable servings per day was significantly different among the groups ($p = 0.015$), and the number of servings was higher in rural residents than in general and nutrition students.

Table 2. Sociodemographic characteristics of the participants ($n = 60$).

	Total ($n = 60$)	Rural Residents ($n = 16$)	General Students ($n = 17$)	Nutrition Students ($n = 27$)	p [1]
Age (year)	31.3 (18.6)	59.4 (14.7) [a]	20.2 (0.8) [b]	21.7 (0.7) [b]	<0.001
Women (n)	35 (58.3)	8 (50.0)	5 (29.4) [a]	22 (81.5) [b]	0.002
BMI (kg/m^2)	21.5 (2.5)	23.1 (2.8) [a]	22.5 (2.0) [a]	19.9 (1.7) [b]	<0.001
Living alone (n)	26 (43.3)	0 (0.0) [a]	13 (76.5) [b]	13 (48.1) [b]	<0.001
No. of V servings/day		[a]	[b]	[b]	0.015
Very few (n)	4 (6.7)	0 (0.0)	2 (11.8)	2 (7.4)	
1–2 servings (n)	32 (53.3)	5 (31.3)	10 (58.8)	17 (63.0)	
3–4 servings (n)	21 (35.0)	9 (56.3)	4 (23.5)	8 (29.6)	
5–6 servings (n)	2 (3.3)	1 (6.3)	1 (5.9)	0 (0.0)	
≥7 servings (n)	1 (1.7)	1 (6.3)	0 (0.0)	0 (0.0)	

BMI, body mass index; V, vegetable. Values are mean (standard deviation) or n (%). [1] Analysis of variance for age and BMI, chi-square test for gender and living alone, and Kruskal-Wallis test for number of vegetable servings (different letters in the same row indicate significant differences by Bonferroni correction for multiple comparisons).

The vegetable variety of the participants is shown in Table 3. The vegetable variety for all participants showed a converging trend in the mean (variation) number of different vegetables consumed per number of days: 6.0 items for one day, 9.2 (+3.2) items for two days, 11.8 (+2.6) items for three days, 13.8 (+2.0) items for four days, 15.3 (+1.4) items for five days, 16.7 (+1.4) items for six days, and 17.9 (+1.2) items for seven days. Since there was no interaction effect between time and group, the main effect of group was analyzed independently. The number of vegetables consumed over seven consecutive days was significantly different in repeated measures ($p = 0.013$), and there were more items for rural residents than for general students and nutrition students in multiple comparisons. There were no significant differences between the groups in the capture proportion of the number of vegetables consumed per survey day based on seven consecutive days. Pearson's correlation coefficient for the correlation with the number of vegetables consumed on seven consecutive days was not statistically significant only on the first day for nutrition students (r = 0.21, $p = 0.292$) and was strong after the third day for all groups (r > 0.70, $p < 0.001$).

To calculate the reference values, when the reciprocal of the number of survey days was taken on the horizontal axis and the reciprocal of the mean number of vegetables consumed in each group on the vertical axis, a high correlation was confirmed between them for each group (Pearson's correlation coefficient, all r > 0.99, $p < 0.001$). The maximum theoretical number of vegetables consumed was 24.8 items for all participants, 27.0 items for rural residents, 24.8 items for general students, and 24.0 items for nutrition students. The capture proportion of the number of vegetables consumed for seven consecutive days calculated from the maximum theoretical number was 71.9% for all participants, 76.3% for rural residents, 69.4% for general students and 69.2% for nutrition students.

Table 3. Vegetable variety of the participants ($n = 60$).

	Total (*n* = 60)	Rural Residents (*n* = 16)	General Students (*n* = 17)	Nutrition Students (*n* = 27)	*p* [1]
No. of V consumed		[a]	[b]	[b]	0.013
1 days	6.0 (3.4)	7.9 (3.9)	5.3 (3.9)	5.4 (2.4)	
2 days	9.2 (4.3)	12.1 (3.7)	8.4 (4.1)	8.0 (4.0)	
3 days	11.8 (4.5)	14.4 (4.5)	10.5 (4.0)	11.1 (4.3)	
4 days	13.8 (4.5)	16.5 (4.5)	12.7 (4.2)	12.9 (4.2)	
5 days	15.3 (5.0)	18.3 (5.6)	14.1 (4.7)	14.2 (4.2)	
6 days	16.7 (5.4)	19.5 (5.7)	15.9 (5.8)	15.4 (4.3)	
7 days (reference value)	17.9 (5.4)	20.6 (5.9)	17.2 (5.6)	16.6 (4.5)	
Max theoretical No.	24.8	27.0	24.8	24.0	
Capture proportion					
1 days/7 days	33.5 (15.6)	37.2 (12.7)	29.2 (18.2)	34.0 (15.3)	0.334
2 days/7 days	51.3 (17.1)	59.3 (10.9)	48.0 (17.5)	48.6 (18.8)	0.091
3 days/7 days	66.0 (13.9)	70.2 (11.2)	61.8 (14.0)	66.2 (14.8)	0.221
4 days/7 days	77.5 (11.5)	80.8 (9.1)	74.7 (11.8)	77.4 (12.4)	0.315
5 days/7 days	85.6 (9.6)	88.5 (6.8)	82.4 (11.3)	85.9 (9.7)	0.190
6 days/7 days (reference value)	93.1 (7.3)	94.6 (6.1)	92.0 (9.3)	93.0 (6.6)	0.609
7 days/Max days [2]	71.9	76.3	69.4	69.2	
Correlation coefficient					
1 days–7 days	0.62 ***	0.78 ***	0.69 **	0.21	
2 days–7 days	0.72 ***	0.89 ***	0.72 **	0.51 **	
3 days–7 days	0.84 ***	0.93 ***	0.79 ***	0.78 ***	
4 days–7 days	0.91 ***	0.95 ***	0.91 ***	0.87 ***	
5 days–7 days	0.94 ***	0.98 ***	0.90 ***	0.92 ***	
6 days–7 days	0.98 ***	0.98 ***	0.98 ***	0.97 ***	

V, vegetables. Values are the mean (standard deviation), reference value, or Pearson's correlation coefficient (** $p < 0.01$, *** $p < 0.001$). [1] Repeated measures analysis of variance for the number of vegetables consumed among groups (different letters in the same row indicate significant differences by Bonferroni correction for multiple comparisons). [2] Maximum theoretical number of days.

4. Discussion

Vegetable variety based on dietary records for seven consecutive days confirmed the differences between groups by repeated measurements. In this study, there was no significant group difference in the capture proportion per survey day based on seven consecutive days, regardless of the number of vegetables consumed between the groups. Furthermore, there was a significant correlation between the number of vegetables consumed on seven consecutive days and that consumed on two or more days in all groups (r > 0.50), which was especially strong on three or more days (r > 0.70). A few previous studies have used a 24-h recall over two days to determine vegetable variety [24,25]. The present study also suggested that a dietary survey of two or more days would provide an estimate of habitual vegetable variety. Incidentally, because significant correlations between the number of vegetables consumed over seven consecutive days and that consumed on two or more days were noted for participants who included weekends and only weekdays in the first two days, respectively (data not shown), the difference between weekdays and weekends was considered minimal.

Comparison between groups showed that vegetable variety over seven consecutive days was significantly higher in rural residents than in nutrition students. Some previous studies have reported that vegetable variety is associated with household income [26], education [27], social class [27], home ownership [27], and marital status [28]. One of the characteristics of the rural residents compared to students in this study was the low proportion of individuals living alone. Because it has been suggested that family or shared meal frequency is also associated with healthier dietary outcomes [29], the association between living with others and vegetable variety may be important. Furthermore, rural residents had the highest Pearson's correlation coefficient between one and seven days in the present study, suggesting that regular eating habits may be related to increased vegetable variety. Although detailed analysis could not be conducted due to the limited number of participants in this study, the differences between groups may be influenced by each confounding factor. Therefore, future analysis based on factors related to vegetable variety is necessary. Furthermore, as previously reported [19–21], the group with higher vegetable consumption had a greater vegetable variety. Although the vegetable items were adopted from the food numbers in the Standards Tables of Food Composition in Japan-2015-(Seventh Revised Edition) [18], the Guidelines for Measuring Household and Individual Dietary Diversity by the Food and Agriculture Organization of the United Nations (FAO) [30] classify vegetable groups differently into categories such as vitamin A-rich vegetables and tubers, dark green leafy vegetables, and other vegetables. Because dietary diversity scores need to be adapted according to local contexts [30], it is also necessary to consider the habitual consumption of vegetable group variety in Japan.

Limitation

This study had several limitations. The first limitation was that the survey was conducted only at a single point in time. According to the FAO guidelines [30], dietary diversity should be measured during different seasons for a more complete assessment of usual diet in rural communities. Since it has been suggested that there are seasonal differences in the amount of vegetable consumption [31], the capture proportion of the number of vegetables consumed per survey day may also be affected by season. Second, the number of survey days was not large enough. Because the capture proportion of the number of vegetables consumed for seven consecutive days calculated from the maximum theoretical number was approximately 70%, the habitual understanding was considered achieved. However, it should be noted that the limited number of survey days requires a large extrapolation, which may have resulted in errors in the maximum theoretical number. Third, because participants were recruited indirectly in several areas, the cooperation rate is unclear. Depending on the cooperation rate, the participants may have been more health conscious than the population. Fourth, if the participant did not provide the food name of the dish, vegetable items were selected from common recipes on the web. Therefore, some of the vegetable items may have been different from what the participants actually ate. There are limitations in understanding the consumption of commercial products because 33.6% of adults eat out at least once a week, and 45.6% of adults eat take-out lunches or prepared foods at least once a week according to the 2019 National Health and Nutrition Survey, Japan [13]. However, generalization may be accepted because vegetable items with a high frequency of use were adopted. Finally, the survey used qualitative dietary records. Because the amount of consumption was not investigated, no cut-off point has been established, and even the consumption of a small amount was categorized as one item. In a previous study that counted the items consumed in separate mixed dishes as main components [32], an ingredient in mixed dishes was assigned if it contributed at least 10% of the dish's total weight or was listed among the top five components. Counting the number of vegetables consumed considering weight is needed in future studies.

5. Conclusions

Based on seven consecutive days, there were no significant group differences in the capture proportion of vegetables consumed on each survey day, and all groups showed a significant correlation with the number of vegetables consumed on two or more days. The present results suggested that a dietary survey of two or more days provides an estimate of habitual vegetable variety. Considering the habitual understanding of grouped vegetable variety is also needed in future studies.

Author Contributions: Conceptualization, O.K.; methodology, R.K. and O.K.; validation, O.K.; formal analysis, R.K. and O.K.; investigation, O.K.; resources, O.K.; data curation, R.K. and O.K.; writing—original draft preparation, R.K.; writing—review and editing, O.K.; visualization, R.K. and O.K.; supervision, O.K.; project administration, O.K.; funding acquisition, O.K. All authors have read and agreed to the published version of the manuscript.

Funding: This research was funded by JSPS KAKENHI, grant number JP19K14044.

Institutional Review Board Statement: The study was conducted according to the guidelines of the Declaration of Helsinki, and approved by the Institutional Ethics Committee of Kio University (protocol code: H30-31; date of approval: 5 September 2018).

Informed Consent Statement: Informed consent was obtained from all subjects involved in the study.

Data Availability Statement: The datasets generated and analyzed during the current study are not publicly available due to privacy and ethical restrictions but are available from the corresponding author on reasonable request.

Acknowledgments: The authors thank the participants in this study. The authors also wish to thank Nozomi Ishiguro, Hinako Otani, Misaki Kamata, Chiharu Shioyama, Miho Takahashi, Nahoko Tani, and Ayaka Bekku (Kio University) for helping with the investigation; Ayaka Oyama (University of Shizuoka) for helping with the data curation; and Hisayoshi Hayashi (University of Shizuoka) for useful discussions.

Conflicts of Interest: The authors declare no conflict of interest.

References

1. Wang, X.; Ouyang, Y.; Liu, J.; Zhu, M.; Zhao, G.; Bao, W.; Hu, F.B. Fruit and vegetable consumption and mortality from all causes, cardiovascular disease, and cancer: Systematic review and dose-response meta-analysis of prospective cohort studies. *BMJ* **2014**, *349*, g4490. [CrossRef] [PubMed]
2. Aune, D.; Giovannucci, E.; Boffetta, P.; Fadnes, L.T.; Keum, N.; Norat, T.; Greenwood, D.C.; Riboli, E.; Vatten, L.J.; Tonstad, S. Fruit and vegetable intake and the risk of cardiovascular disease, total cancer and all-cause mortality-a systematic review and dose-response meta-analysis of prospective studies. *Int. J. Epidemiol.* **2017**, *46*, 1029–1056. [CrossRef] [PubMed]
3. Wallace, T.C.; Bailey, R.L.; Blumberg, J.B.; Burton-Freeman, B.; Chen, C.-Y.O.; Crowe-White, K.M.; Drewnowski, A.; Hooshmand, S.; Johnson, E.; Lewis, R.; et al. Fruits, vegetables, and health: A comprehensive narrative, umbrella review of the science and recommendations for enhanced public policy to improve intake. *Crit. Rev. Food Sci. Nutr.* **2020**, *60*, 2174–2211. [CrossRef] [PubMed]
4. World Health Organization. *Global Strategy on Diet, Physical Activity and Health*; World Health Organization: Geneva, Switzerland, 2004.
5. Keim, N.L.; Forester, S.M.; Lyly, M.; Aaron, G.J.; Townsend, M.S. Vegetable variety is a key to improved diet quality in low-income women in California. *J. Acad. Nutr. Diet.* **2014**, *114*, 430–435. [CrossRef] [PubMed]
6. Ye, X.; Bhupathiraju, S.N.; Tucker, K.L. Variety in fruit and vegetable intake and cognitive function in middle-aged and older Puerto Rican adults. *Br. J. Nutr.* **2013**, *109*, 503–510. [CrossRef]
7. Cooper, A.J.; Sharp, S.J.; Lentjes, M.A.H.; Luben, R.N.; Khaw, K.-T.; Wareham, N.J.; Forouhi, N.G. A prospective study of the association between quantity and variety of fruit and vegetable intake and incident type 2 diabetes. *Diabetes Care* **2012**, *35*, 1293–1300. [CrossRef] [PubMed]
8. Büchner, F.L.; Bueno-de-Mesquita, H.B.; Ros, M.M.; Overvad, K.; Dahm, C.C.; Hansen, L.; Tjønneland, A.; Clavel-Chapelon, F.; Boutron-Ruault, M.-C.; Touillaud, M.; et al. Variety in fruit and vegetable consumption and the risk of lung cancer in the European Prospective Investigation into Cancer and Nutrition. *Cancer Epidemiol. Biomarkers Prev.* **2010**, *19*, 2278–2286. [CrossRef] [PubMed]
9. U.S. Department of Agriculture and U.S. Department of Health and Human Services. *Dietary Guidelines for Americans 2020–2025*; USDA Publication and HHS Publication: Washington, DC, USA, 2020.
10. National Health and Medical Research Council. *Australian Dietary Guidelines*; National Health and Medical Research Council: Canberra, Australia, 2013.
11. Marshall, A.N.; van den Berg, A.; Ranjit, N.; Hoelscher, D.M. A scoping review of the operationalization of fruit and vegetable variety. *Nutrients* **2020**, *12*, 2868. [CrossRef] [PubMed]
12. Willett, W. Food frequency methods. In *Nutritional Epidemiology*, 3rd ed.; Willett, W.C., Ed.; Oxford University Press: New York, NY, USA, 2013; pp. 70–95.
13. Ministry of Health, Labour and Welfare. *The National Health and Nutrition Survey in Japan.* 2019. Available online: https://www.mhlw.go.jp/content/000710991.pdf (accessed on 10 November 2021).
14. Drewnowski, A.; Henderson, S.A.; Driscoll, A.; Rolls, B.J. The dietary variety score: Assessing diet quality in healthy young and older adults. *J. Am. Diet. Assoc.* **1997**, *97*, 266–271. [CrossRef]

15. Asato, L.; Hiroi, Y.; Shirota, T.; Toyokawa, H.; Shinjo, S.; Yamamto, S. Analyses of food intake by female college students and their dietary habits: Saturation curve of number of consumed food items accompanying increase of number of meals. *Jpn. J. Nutr. Diet.* **1992**, *50*, 275–283. [CrossRef]
16. Ministry of Agriculture, Forestry and Fisheries. 2015 Census of Agriculture and Forestry in Japan Report and Data on the Results. Available online: https://www.maff.go.jp/j/tokei/census/afc2015/280624.html (accessed on 10 November 2021).
17. Baranowski, T. 24-hour recall and diet record methods. In *Nutritional Epidemiology*, 3rd ed.; Willett, W.C., Ed.; Oxford University Press: New York, NY, USA, 2013; pp. 49–69.
18. Ministry of Education, Culture, Sports, Science and Technology. Standards Tables of Food Composition in Japan-2015-(Seventh Revised Edition). Available online: https://www.mext.go.jp/en/policy/science_technology/policy/title01/detail01/1374030.htm (accessed on 10 November 2021).
19. Meengs, J.S.; Roe, L.S.; Rolls, B.J. Vegetable variety: An effective strategy to increase vegetable intake in adults. *J. Acad. Nutr. Diet.* **2012**, *112*, 1211–1215. [CrossRef] [PubMed]
20. Sim, M.; Blekkenhorst, L.C.; Lewis, J.R.; Bondonno, C.P.; Devine, A.; Zhu, K.; Woodman, R.J.; Prince, R.L.; Hodgson, J.M. Vegetable diversity, injurious falls, and fracture risk in older women: A prospective cohort study. *Nutrients* **2018**, *10*, 1081. [CrossRef] [PubMed]
21. Ashton, L.; Williams, R.; Wood, L.; Schumacher, T.; Burrows, T.; Rollo, M.; Pezdirc, K.; Callister, R.; Collins, C.E. The comparative validity of a brief diet screening tool for adults: The Fruit and Vegetable VAriety index (FAVVA). *Clin. Nutr. ESPEN* **2019**, *29*, 189–197. [CrossRef]
22. Kushida, O.; Murayama, N.; Iriyama, Y.; Horikoshi, K.; Takemi, Y.; Yoshiike, N. An algorithm for assessing changes in vegetable intake behavior in adult male in Japan. *Jpn. J. Nutr. Diet.* **2011**, *69*, 294–303. [CrossRef]
23. Dowd, J.E.; Riggs, D.S. A comparison of estimates of michaelis-menten kinetic constants from various linear transformations. *J. Biol. Chem.* **1965**, *240*, 863–869. [CrossRef]
24. Azadbakht, L.; Mirmiran, P.; Azizi, F. Variety scores of food groups contribute to the specific nutrient adequacy in Tehranian men. *Eur. J. Clin. Nutr.* **2005**, *59*, 1233–1240. [CrossRef] [PubMed]
25. Vandevijvere, S.; De Vriese, S.; Huybrechts, I.; Moreau, M.; Van Oyen, H. Overall and within-food group diversity are associated with dietary quality in Belgium. *Public Health Nutr.* **2010**, *13*, 1965–1973. [CrossRef]
26. Giskes, K.; Turrell, G.; Patterson, C.; Newman, B. Socio-economic differences in fruit and vegetable consumption among Australian adolescents and adults. *Public Health Nutr.* **2002**, *5*, 663–669. [CrossRef]
27. Conklin, A.I.; Forouhi, N.G.; Suhrcke, M.; Surtees, P.; Wareham, N.J.; Monsivais, P. Variety more than quantity of fruit and vegetable intake varies by socioeconomic status and financial hardship. Findings from older adults in the EPIC cohort. *Appetite* **2014**, *83*, 248–255. [CrossRef] [PubMed]
28. Vinther, J.L.; Conklin, A.I.; Wareham, N.J.; Monsivais, P. Marital transitions and associated changes in fruit and vegetable intake: Findings from the population-based prospective EPIC-Norfolk cohort, UK. *Soc. Sci. Med.* **2016**, *157*, 120–126. [CrossRef] [PubMed]
29. Fulkerson, J.A.; Larson, N.; Horning, M.; Neumark-Sztainer, D. A review of associations between family or shared meal frequency and dietary and weight status outcomes across the lifespan. *J. Nutr. Educ. Behav.* **2014**, *46*, 2–19. [CrossRef]
30. Kennedy, G.; Ballard, T.; Dop, M.C. *Guidelines for Measuring Household and Individual Dietary Diversity*; Food and Agriculture Organization of the United Nation: Rome, Italy, 2013.
31. Sasaki, S.; Takahashi, T.; Iitoi, Y.; Iwase, Y.; Kobayashi, M.; Ishihara, J.; Akabane, M.; Tsugane, S. Food and nutrient intakes assessed with dietary records for the validation study of a self-administered food frequency questionnaire in JPHC Study Cohort I. *J. Epidemiol.* **2003**, *13*, 23–50. [CrossRef] [PubMed]
32. Conklin, A.I.; Monsivais, P.; Khaw, K.T.; Wareham, N.J.; Forouhi, N.G. Dietary diversity, diet cost, and incidence of type 2 diabetes in the United Kingdom: A prospective cohort study. *PLoS Med.* **2016**, *13*, e1002123. [CrossRef] [PubMed]

Article

Development and Validation of the Vietnamese Children's Short Dietary Questionnaire to Evaluate Food Groups Intakes and Dietary Practices among 9–11-Year-Olds Children in Urban Vietnam

Thi My Thien Mai [1,2,*], Quoc Cuong Tran [3], Smita Nambiar [1], Jolieke C. Van der Pols [1] and Danielle Gallegos [1,4]

[1] School of Exercise and Nutrition Sciences, Queensland University of Technology (QUT), Victoria Park Rd., Kelvin Grove, QLD 4059, Australia
[2] Ho Chi Minh City Center for Disease Control, Ho Chi Minh City 700000, Vietnam
[3] Department of Nutrition and Food Safety, Faculty of Public Health, Pham Ngoc Thach Medical University, Ho Chi Minh City 700000, Vietnam
[4] Woolworths Centre for Childhood Nutrition Research, Queensland University of Technology (QUT), Graham St., South Brisbane, QLD 4101, Australia
*   Correspondence: thimythien.mai@hdr.qut.edu.au

Abstract: This study aims to develop and assess the reproducibility and validity of the Vietnamese Children's Short Dietary Questionnaire (VCSDQ) in evaluating food groups intakes and dietary practices among school-aged children 9–11 years old in urban Vietnam. A 26-item questionnaire covering frequency intakes of five core food groups, five non-core food groups, five dietary practices over a week, and daily intakes of fruits, vegetables, and water was developed. Children ($n = 144$) from four primary schools in four areas of Ho Chi Minh City, Vietnam completed the VCSDQ twice, as well as three consecutive 24 h recalls over a week. Intra-class correlation, Spearman correlation, weighted kappa, cross-classification, and Bland–Altman plots were used to evaluate the reproducibility and validity. The direct validity of food groups from VCSDQ against the 24 h recalls was examined using Wilcoxon-test for trend. The VCSDQ had good reproducibility in 12 out of 15 group items; the ICC ranged from 0.33 (grains) to 0.84 (eating while watching screens). This VCSDQ had low relative validity, two items (instant noodles, eating while watching screens) had a moderate to good agreement (k = 0.43, k = 0.84). There was good direct validity in three core-food groups (fruits, vegetables, dairy) and three non-core food groups (sweetened beverages, instant noodles, processed meat). In addition, the VCSDQ can also be used to classify daily intakes of fruits and vegetables from low to high.

Keywords: validation; short dietary questionnaire; food groups; dietary practices; children; Vietnam

Citation: Mai, T.M.T.; Tran, Q.C.; Nambiar, S.; Pols, J.C.V.d.; Gallegos, D. Development and Validation of the Vietnamese Children's Short Dietary Questionnaire to Evaluate Food Groups Intakes and Dietary Practices among 9–11-Year-Olds Children in Urban Vietnam. *Nutrients* **2022**, *14*, 3996. https://doi.org/10.3390/nu14193996

Academic Editors: Colin Bell and Penny Love

Received: 15 August 2022
Accepted: 21 September 2022
Published: 26 September 2022

Publisher's Note: MDPI stays neutral with regard to jurisdictional claims in published maps and institutional affiliations.

1. Introduction

Childhood overweight and obesity is a global public health issue affecting more than 340 million children aged 5–19 years [1]. The most significant rise in the prevalence of childhood overweight and obesity has been in low-middle income countries [2]. Childhood overweight and obesity is responsible for short and long-term impacts on children's health, including increased risk of cardiovascular diseases, diabetes, cancer, psychological and mental disorders, and increases the probability of becoming an adult with overweight and obesity [2,3]. In Vietnam, a lower-middle income country in Asia, the prevalence of overweight and obesity in children 5–19 years old has doubled from 8.5% in 2010 to 19% in 2019, with nearly 30% of children living in urban areas being overweight or obese [4]. Notably, in Ho Chi Minh City, the largest city in Vietnam, the prevalence of overweight and obesity in primary school children in 2014 was 51%. Of these, 27% of children were obese, doubling every five years since 2004 [4]. Childhood overweight and obesity is, therefore, a serious public health issue in urban Vietnam. The rise in childhood overweight and obesity has coincided with national economic growth and the co-occurring nutrition transition with rapid changes towards a less healthful diet [5]. This has been primarily the case in urban Vietnam, where this economic growth is characterized by increased intakes of ultra-processed foods and eating outside of the home in emerging multinational fast-food chains [5]. The rapidity of the changes in diet and food environments has been difficult to monitor, making timely public health action challenging.

Promoting a healthy diet is one strategy to prevent the increase in overweight and obesity. However, there is little information on food intake and dietary practices among Vietnamese children. Current dietary assessment in national surveys uses 24 h recalls, which are costly, time-consuming, and tend to focus on nutrients rather than food intake and dietary patterns. In addition, the Vietnamese national nutrition survey is only conducted every ten years and has therefore not been able to keep up to date with emerging trends [6]. These factors limit the opportunity to provide salient data regarding children's food intake and dietary practices in the context of overweight and obesity that could inform public health policy and strategies. Thus, dietary assessment tools that are low cost, easy to administer and able to provide in-time data about children's food intakes and dietary practices are needed.

Short dietary questions, used individually or together as a questionnaire, have previously been used as a tool to evaluate and monitor population food intakes and compliance with dietary guidelines [7,8]. Most short dietary questions focus on specific food groups such as fruits and vegetables, dairy groups, or sweetened beverages, as well as some dietary practices such as meal skipping rather than taking a whole diet approach [9]. In addition, most short dietary questions have been developed and validated among children in high-income countries where the food context (environment and intake) differs from those in lower-middle-income countries. Consequently, our primary aim was to develop a short dietary questionnaire to evaluate usual food group intakes and dietary practices over a one-week period for children aged 9–11 years old. The secondary aim was to examine the relative validity and reproducibility of the newly developed short dietary questionnaire against three 24 h food recalls among fifth-grade students in urban Vietnam.

2. Materials and Methods

This study follows the Best Practices for Conducting and Interpreting Studies to Validate Self-Report Dietary Assessment Method, which is partly adapted from the checklist for nutritional epidemiology study (STROBE-nut) [10]. See Supplementary Materials Table S1 for the STROBE-nut checklist.

2.1. Study Design

A validation study design was used to examine the reproducibility and relative validity of the Vietnamese Children's Short Dietary Questionnaire (VCSDQ) among children aged 9–11 years in Ho Chi Minh City. The VCSDQ was validated against three non-consecutive 24 h recalls collected within one week (two weekdays and a weekend). The reproducibility of the VCSDQ was tested using repeated administration one week apart.

2.2. Setting

This study occurred at primary schools located in urban and rural areas within Ho Chi Minh City, Vietnam. Ho Chi Minh City is the most populous city in Vietnam, with over nine million people, population density and economic development [11,12]. Due to disruptions caused by the COVID-19 pandemic, data was collected in two periods (July 2020 and September–October 2020).

2.3. Participants

A sample size of approximately 160 was required to examine the correlations of intake frequencies between the VCSDQ, and the repeated 24 h recalls ($p < 0.05$, 80% power) [13,14]. A response rate of about 50% was expected based on a previous study [15], consequently, a minimum of 320 students were invited to participate in the research.

Grade-five students were recruited because students in this age group (9–11 years old) have been identified to be able to report their diet using a food frequency questionnaire more accurately than their parents [16]. In Vietnam, primary school starts at Grade 1 when children are 6–7 years old, and secondary school starts at Grade 6 when they are 11–12.

Multi-stage sampling was applied to select participating schools. Firstly, based on geographical categorizations, the city was divided into four areas: four wealthy urban districts, nine less wealthy urban districts, six emerging urban districts, and five rural districts. Four districts were randomly selected from each of these four areas. Then, three primary schools (one for selection and two for backup) per district were randomly selected ($n = 48$). After approaching four selected schools, three agreed to participate in the study. Another school from the list of backup schools in the same district was invited to participate in the study. To achieve an equal sample of student participants from each school ($n = 40$), two to three grade five classes were randomly selected to participate based on the school size.

Information and consent forms were given to 447 children and sent home to the parents/caregivers. Only 163 children who returned two completed consent forms (one from the parent and one from the child) within a week participated in the research. Exclusions included children with cognitive impairments who were unable to complete the survey independently and children with scoliosis or other musculoskeletal disorders or who were restricted to a wheelchair due to the inability to collect accurate anthropometric data.

2.4. Development of a Vietnamese Children's Short Dietary Questionnaire (VCSDQ)

The VCSDQ was developed in English by TMTM, based on the Vietnamese food-based dietary guidelines to cover food groups and dietary practices [17]; the Food Pyramid for children aged 6–11 years old for recommendations of vegetables, fruit serving sizes and water intake [18]; and by reviewing available short dietary questions for children aged 9–11 years old [7,8], internationally to identify common items and structure of short dietary questions. The first draft of the VCSDQ consisted of 29 open- and closed-ended questions on the frequency and serves of food groups intake and mealtime behaviors over the last week. This draft was then translated into Vietnamese and evaluated for content validity by five nutrition experts (local Vietnamese academics with experience in dietary data collection with children). Experts were asked to rate the relevance and clarity of each item using a four-point ordinal scale: 1 = not relevant/clear, 2 = somewhat relevant/clear, 3 = quite relevant/clear, 4 = highly relevant/clear [19]. Three out of five experts rated the VCSDQ as 3, and two rated the questionnaire as 4. The suggestions from experts rated as 3 included improvements in the clarity of languages suitable for the age of primary schools and for Southern dialects as Ho Chi Minh City is in the south of Vietnam. In addition, it was recommended that food items should be re-arranged according to the food groups

After this process, modifications to the first draft were made, and the second draft of the VCSDQ was reviewed with three children aged 9–11 years old. Cognitive interviewing was used to assess how children understood the questions and their process in answering questions and to identify issues related to understanding the questions and responses. The cognitive interview was recorded, and the main issues were coded using four categories: understanding the question, difficulty related to recalling information, problems with identifying the frequency of consumption or portion size, and difficulty with selecting answer options. Most students found the questionnaire easy to understand but did have difficulty in identifying serves of foods with open-ended questions. Consequently, the revised version of VCSDQ also included three closed questions about the average number of serves of fruits, vegetables and the amount of water consumed daily over the last week, with pictures below each question to illustrate one serve of fruits/vegetables and a cup of water. In addition, to facilitate timely completion and analysis, the paper version of the VCSDQ was designed as an online version using Key Survey™ (Version 8.70, WorldAPP, Quincy, MA, USA) [20]. This version was then translated to English and reviewed by two authors (DG, SNM), and the Vietnamese version was then tested with six students to evaluate readability and the time taken to complete the VCSDQ. All students could answer the VCSDQ by themselves, taking between 6 to 15 min to complete.

The final VCSDQ consisted of 26 items in two languages English and Vietnamese. The first 17 items asked about the frequency of intake of five core food groups (grains, vegetables, fruits, meat and alternatives, dairy) and five non-core food groups (sugar-sweetened beverages, sweets and savory snack foods, fast food, instant noodles, processed meats). The following section included six items about the frequency of five dietary practices (adding salt/sauces, eating a meal with an adult in the family, eating while watching television and other screens, eating a meal cooked outside the home, and skipping meals). The final section was three items about daily intake of fruits, vegetables, and water in serves/cups. Each food group item contained eight response options: not eaten, less than once/week, once/week, 2–4 times/week, 5–6 times/week, one/day, 2–3 times/day, 4 times/day or more. For the six items related to dietary practices, the response options "less than once/week" was removed after cognitive interviewing with children as it was poorly understood. For questions about daily intakes, answer options were added: not eating, less than once serve, once serve/day, 2 serves/day, 3 serves/day, 4 serves/day or more for intake of fruits and vegetables, and options: not drinking water, one cup/day or less, 2–3 cups/day, 4–5 cups/day, 6–7 cups/day, 8 cups/day or more for water intake. In addition, to facilitate the estimation of serves of fruits and vegetables or cups of water, pictures with the example serving sizes of fruit and vegetables and a picture with a cup and bottle of water, were added below the question to increase comprehension and help improve the accuracy of estimation. In addition, to facilitate familiarity with the questionnaires, instructions for answering the questions and two practice questions were added prior to the actual VCSDQ questions. See Figure S1 for a copy of the VCSDQ.

2.5. Reference Method: 24 h Recalls over Three Separate Days

Children were asked to recall what they ate and drank 24 h before the interview, using the 5-step multiple-pass method developed by the USDA [14]. Each child was interviewed at school three times in one week to collect 24 h recalls from two weekdays and one weekend day. To improve the estimation of the foods and drinks consumed, a book with food pictures showing portion sizes and everyday utensils to help understand portion sizes were used. In addition, to improve the quality of the 24 h recalls, the school menu with the actual picture of one serving and the ingredients were collected before the interview to support interviewers and facilitate children in reporting and estimating their meals consumed at school. All data from the 24 h recalls were recorded on a prepared 24 h data collection sheet by interviewers trained in 24 h recall methods by the lead investigator. Due to the limited time allocation from school and the availability of interviewers during each data collection period, 20 interviewers who were students from the Nutrition and Dietetics and Preventive Medicine programs at the Ho Chi Minh City University of Medicine and Pharmacy were recruited to participate in a two-day training workshop for data collection. After this workshop, each interviewer was asked to submit an actual record of a 24 h recall to be reviewed and provided with suggestions for improvements until the record accurately represented the previous day's intake. In each period, 8–12 interviewers were invited to collect the 24 h recalls. In addition, two health officers who were experienced with 24 h recall interviews checked for any unusual food intakes or missed food records.

Each 24 h recall record was converted into grams of food intake or mL of beverage consumption using a food weight conversion table. In addition, a book of pictures of common foods for dietary assessment in children developed by the National Institute of Nutrition [21], a book of street foods with usual portion size in Ho Chi Minh City developed by Ho Chi Minh City Nutrition Centre [22], and unpublished conversion databases from Ho Chi Minh City Nutrition Center [23] were used for the estimates. After the first round of data entry, the conversion database of ten food items was unavailable, so the researchers weighed two samples of the food item, and average values were used to determine its weight.

Intake amounts of all reported foods and drinks were entered into a Microsoft Excel Spreadsheet and assigned a food code using the 2017 Vietnamese food composition tables [24]. If food-items were not available in the Vietnamese food composition tables for example, chia seeds or oats, a new food-code was created and the nutrient composition was borrowed from either the ASEAN Food Composition Table, USDA food composition data. In addition, food composition data from countries closest to Vietnam such as Singapore, Thailand, and India or the country from where the food was exported to Vietnam such as Australia or Japan were used for reference. If the food was not available in other food composition tables, then the data was inputted using nutrients reported on the food label if available or was calculated based on the food composition of ingredients for composite foods (e.g., dried chicken with lemon leaf, octopus' ball, Oreo™ cookies and milk blended with ice).

Each record included: study ID, date of visit, name of the dish, time of eating, name of a meal (breakfast/lunch/dinner/suppers/snack), place of eating (school/home/shop/on the way to school or from school to home), home-cooked (yes/no), food code, name of food, the weight of each food, amount of nutrients, start and end time of the interview, duration of the interview, whether a child was consuming their usual diet or not, whether the child was following a special diet or not. In addition, information about whether children ate meals (breakfast, lunch, dinner) with family members or watched screens at these meals during the 24 h recall period was recorded to facilitate the validation of the VCSDQ.

2.6. Matching Food Groups and Dietary Practices from the 24 h Recall with the VCSDQ

To facilitate the comparison between the VCSDQ and the 24 h recalls, all data from the 24 h recalls, and the VCSDQ were converted into frequencies per day for the 17 items of frequency intakes and the six of dietary practices items.

For the VCSDQ, the frequencies per day was converted using the conversion table (Table 1):

Table 1. Conversion table compiled based on the calculation from [25].

VCSDQ Response	Frequency	Calculation	Times/Day
1	Never/not eating	0	0.00
2	less than once/week	0.5/7	0.07
3	once/week	1.0/7	0.14
4	2–4 times/week	3.0/7	0.43
5	5–6 times/week	5.5/7	0.79
6	once/day	1.0/1	1.00
7	2–3 times/day	2.5/1	2.50
8	≥4 times/day	4/1	4.00

For the 24 h recalls, the frequencies per day were calculated using the following steps.

Firstly, the frequency intakes of each food group for each 24 h recall were calculated based on the number of eating occasions. Each eating occasion was defined as all food and drink consumed within a 30 min interval [26]. For example, if vegetables were consumed on two eating occasions across 24 h, the number of vegetable intakes over 24 h were counted as twice. However, with two different types of vegetables consumed on one eating occasion, the number of vegetable intakes was still counted as once [26]. All primary ingredients from the 24 h recalls were allocated into one of the 17 food items (five core and five non-core food groups) as reported in the VCSDQ by the lead author (TMTM) and reviewed by other authors (DG, SMN, JVDP) using the food classifications from the Vietnamese Food Composition Table [24]. The number of food groups intakes over 24 h was calculated by the total number of intakes of those food groups for each eating occasion.

For dietary practices, the number of times the child ate with other family members and watched screens while eating during breakfast, lunch or dinner were calculated from the same question in the 24 h recalls. Other dietary practices, including adding salt, adding sauces while eating, eating outside the home, and skipping meals, were defined, and coded based on the information from the 24 h recalls. Adding salt was defined as the number of eating occasions that the child added salt while eating fruit (it is a cultural practice to eat fruits with salt instead of using table salt to add to foods in western countries) or adding fish or soya sauce while eating other dishes. Adding sauces was defined as the number of eating occasions that children added other sauces such as tomato sauces, chilli sauces, and mayonnaise while eating. Eating outside of the home was defined as the number of eating occasions that cooked foods were consumed at shops or restaurants or home but were not homemade. Skipping meals was defined as the number of times children did not eat anything for breakfast, lunch or dinner.

Secondly, as the VCSDQ asked about the frequency of intakes/eating habits over the last week, the daily number/weight of intakes/dietary practices from the 24 h recalls were adjusted for a whole week using the following equation: average of intakes from two weekdays multiplied by five plus intakes on the weekend multiplied by two and this total of a week's intake was divided by seven. Ten children had only two 24 h recalls; a third record was created using the average value of the other two records prior to applying this equation. If a child had only one 24 h recall, they were excluded, but there were no cases with one record. In addition, the daily intake of fruits, vegetables, and water from 24 h recalls was calculated to facilitate the comparison with three items from the SDQ (mean daily intake of fruits, vegetables, and water over the last week).

2.7. Anthropometric Measurements

All participating children had weight, height, and waist circumference measured twice by trained health officers using standard measurement protocols [27]. If the difference between two measurements for height was 0.5 cm, for weight 0.5 kg, and waist circumference 1 cm or more, a third measurement was conducted. Height was measured by using a wooden height board to the nearest 0.1 cm, weight was measured using scales (TANITA HD-318, TANITA, Tokyo, Japan) to the nearest 0.1 kg, waist circumference was measured by non-plastic tapes to the nearest 0.1 cm, at the end of normal expiration at the midpoint between the lower margin of the last palpable rib and the top of the iliac crest.

Covariates

Child characteristics: sex and date of birth were recorded during the anthropometric assessment.

Overweight and obesity: data from the anthropometric measurements were used to generate BMI z-scores for age using the WHO Anthropometric macro [28]. According to recommendations from WHO [28], overweight was defined as BMI z-for-age > 1 SD, and obesity as BMI z-for-age > 2 SD.

2.8. Procedures

Data collection took place in a private location at the school suitable for interviewing and the collection of anthropometric measurements over three visits (with one visit on Monday to collect weekend diet data). The first visit and the third visit were one week apart. The 24 h recalls were collected in three visits by the interviewer with children's reports. The VCSDQ was completed by children at the first and third visit with the support from research team members using Key Survey™ App, Version 8.7, WorldAPP, Quincy, MA, USA on tablets. The process of validation study is presented in Figure 1.

2.9. Bias

The food intake records from the average of three 24 h recalls were reviewed to determine under- and over-reporting using Goldberg's method [29,30]. The under- and over-reporting of dietary intake were identified by comparing the ratio of reported energy intake (EIrep) to the predicted basal metabolic rate (BMRest) with the 95% upper and lower confidence limit of physical activity level (PAL):

$$\text{PAL} \times \exp\left[SDmin \times \frac{S/100}{\sqrt{n}} < \text{EIrep} : \text{BMRest} < \text{PAL} \times \exp\left[SDmax \times \frac{S/100}{\sqrt{n}}\right]\right.$$

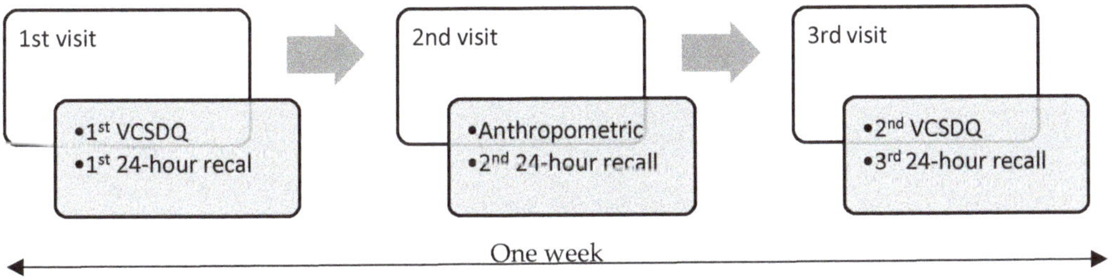

1st visit		2nd visit		3rd visit
•1st VCSDQ •1st 24-hour recal	→	•Anthropometric •2nd 24-hour recall	→	•2nd VCSDQ •3rd 24-hour recall

One week

Figure 1. Process of validation study of Vietnamese Children's Dietary Short Questionnaire.

The mean of energy intake from the three 24-h recalls for each student was used as estimated energy intake. The predicted basal metabolic rate was calculated for each child using established equations adjusted for weight, height, relevant sex, and age groups in consideration the context of overweight and obesity in the study population [31,32]. The 95% lower and upper limit were calculated with the physical activity level was 1.55 due to the evidence of the low level of physical activity among fifth-grade children in Ho Chi Minh City [15] and the average within-subject variation in intake (S) calculated by the equation below, with suggested values for within-subject variation in energy intake (CV_{wEI}), within-subject variation in repeated BMR measurement (CV_{wB}), total between-subject variation in PAL (CV_{tP}), and the number of days of dietary assessments were 23%, 8.5% and 15%, 3 days, respectively [33].

$$S = \sqrt{\frac{CV_{wEI}^2}{d} + CV_{wB}^2 + CV_{tP}^2}$$

2.10. Statistical Analyses

The total frequencies of intake per day for the core and non-core food groups, and the frequency per day of the different assessed dietary practices, were used for the analysis. The distribution of all variables resulting from the VCSDQ and the 24 h recalls were examined. If variables did not have a normal distribution, the median of frequencies rather than the mean was presented. Consumption frequencies per day from VÇSDQ and the 24 h-recalls were converted to quartiles for subsequent analysis.

2.10.1. Reproducıbılity

The average intra-class correlation (ICC) (two-way mixed-effect model, absolute agreement) was used to examine the reproducibility of the VCSDQ after one week. ICC values were categorized as <0.5, 0.5 to <0.75, 0.75 to <0.9, and ≥0.9, indicating that the questionnaire had poor, moderate, good, or excellent reproducibility for the food group analyzed, respectively [34]. In addition, Spearman correlation coefficients, classification of agreement, and weighted kappa were applied to further examine the reproducibility of the VCSDQ.

2.10.2. Relative Validity

The relative validity of VCSDQ was examined by comparing intake frequencies with those derived from the average of the three 24 h recalls as the reference method. Group-level validation tests including Wilcoxon paired signed-rank test, and individual–level validation tests, including Spearman correlation, classification agreement, and weighted kappa to examine the relative validity of the VCSDQ [35,36].

For group-level validation, the equality of the median between the first VCSDQ and 24 h-recalls was examined using Wilcoxon paired signed-rank test as all data were skewed ($p > 0.05$ indicating values from two methods are equal) [35,37]. In addition, Bland-Altman plots were used to evaluate the direction of the bias (under and over-reporting of the VCSDQ compared to three 24 h recalls) at group level by estimating mean differences (VCSDQ minus 24 h recalls) in times per day with 95% limit of agreement (mean $\pm$ 1.96 SD) [35,38].

For individual-level validation, the correlation between VCSDQ and 24 h recalls were firstly examined by Spearman rank correlation coefficient (SCC) to evaluate the strength and direction of the association (>0.5 good outcomes, 0.2–0.49 acceptable outcomes, and <0.2 poor outcomes) [35,36]. Then, to adjust the within-person variation from 24 h-recalls, de-attenuation correlations were calculated using the equation [14]:

$$r_t = r_0 \sqrt{1 + \frac{\lambda_x}{n_x}}$$

where r_t is the de-attenuation correlation, r_0 is the observe correlation, λ_x is the ratio of the within- and between-person variances for the 24 h recalls n_x is the number of repeated measurements of 24 h recalls (n_x = 3).

Cross-classification was used to test whether the child was correctly classified in the same quartile (exact agreement), the same and ± 1 quartile (exact and adjacent agreement) or the opposite quartile (gross misclassification). Good agreement was indicated when the percentage was in the same quartile > 50% or opposite quartile < 10% [35,36]. To further examine the classification agreement, quadratic weighted kappa values were calculated. The strength of the agreement, as defined by Landis and Koch [39], was categorized as value ≤ 0 (no agreement), 0.01–0.20 (slight agreement), 0.21–0.40 (fair agreement), 0.41–0.60 (moderate agreement), 0.61–0.80 (substantial agreement), 0.81–0.99 (perfect agreement).

2.10.3. Direct Validity

Although examining responsiveness to change is generally encouraged in the validation of diet questionnaires, it was not possible to assess change over longer periods in this cross-sectional study. Instead, the direct validity with sensitivity to change was estimated by analyzing mean intakes calculated from 24 h recall across categories of the VCSDQ and examining the trend by applying the Wilcoxon-test for trend, which is an extension of the Wilcoxon rank-sum test [40]. We used this method to assess the existence of any trends in intake of fruits, vegetables, and water from 24 h-recalls across increasing number of serves of intake from the VCSDQ.

All analyses were carried out in STATA statistical software version 17 (StataCorp LLC, College Station, TX, USA). Statistical significance was defined as p-value < 0.05.

2.11. Ethics

The study was conducted under the Declaration of Helsinki and approved by the Queensland University of Technology Human Research Ethics Committee (protocol version 3, approved 30 September 2019; UHREC Reference number: 1900000601).

3. Results

3.1. Participants

One hundred and sixty-three (n = 163) fifth-grade students from four schools in Ho Chi Minh City participated in the study, giving a response rate of 36.5%. Of these, 153 (93.4%) participants completed three 24 h recalls, with the remaining participants completing two recalls. Nineteen (11.6%) participants who misreported (4.3% under-reported and 7.3% over-reported) were excluded from the analysis. After excluding mis-reporters, 144 students were included in the analysis of relative validity. For the analysis of VCSDQ's reproducibility, five students were absent for the second administration of the VCSDQ, so a total of 139 participants were included in the analysis. A summary of participant numbers for each study is presented in Figure 2.

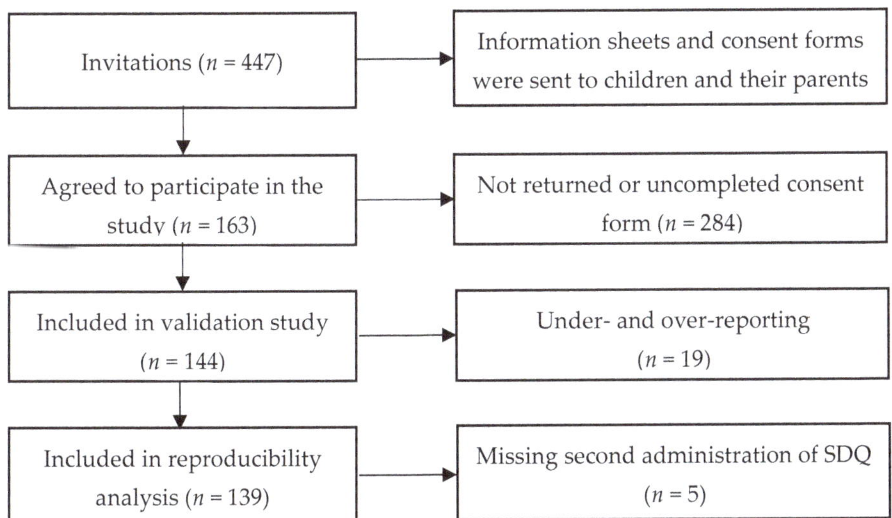

Figure 2. Participants in the validation study.

The characteristics of participants and mis-reporters are presented in Table 2. Boys and girls were evenly represented, and the mean of age was 10.6 ± 0.5 years. Nearly sixty percent (58.9%) of participants were overweight or obese with 31% obese. Median energy intake was 1916 kcal/day, which is higher than in the under-reporter groups (1150 kcal/day) and lower than over-reporters groups (3121 kcal/day). The proportion of children having school lunch among participants was higher than those in the under and over-reporters group. However, there were no significant differences between included participants and those who misreported their dietary intake.

Table 2. Characteristics of participants in the validation of VCSDQ *.

Characteristic	Total Participants (n = 163)	Under and Over Reporters (n = 19)	Participants in Validation Study (n = 144)
Sex, male	81 (49.7)	10 (52.6)	71 (49.3)
Age (year)	10.6 ± 0.5	10.8 ± 0.6	10.6 ± 0.5
Weight (kg)	41.5 (34.1–49.3)	41.5 (33.7–52.7)	41.5 (34.1–48.5)
Height (cm)	143.2 ± 7.1	143.3 ± 7.2	142.9 ± 6.9
BMI z-score (SD)	1.25 (0.2–2.2)	1.3 (0.3–2.2)	1.3 (0.2–1.3)
Nutritional status			
Thinness	5 (3.1)	1 (5.3)	4 (2.8)
Normal	62 (38.0)	8 (42.1)	54 (37.5)
Overweight	46 (28.2)	3 (15.8)	43 (29.9)
Obesity	50 (30.7)	7 (36.8)	43 (29.9)
Energy intake (kcal/day)	1932 (1625–2232)	2581 (1185–3285)	1916 (1645–2178)
Having school lunch	116 (71.2)	10 (52.6)	106 (73.6)

* The data is presented as mean $\pm$ SD or median (25th–75th) or n (%).

3.2. Descriptive Data

Overall, the frequencies of intakes and dietary practices, and daily food intakes between the VCSDQ and 24 h recalls were significantly different across most food groups and dietary practices except for the frequency intakes of fruits, instant noodles, and skipping meals (Table 3). Records from 24 h-recalls indicated that children consumed fruits, dairy products, sweetened beverages, snacks, and discretionary foods, had family meals, and ate out of home at least once per day. They consumed vegetables twice a day and grains, meat, and alternatives three times a day. In contrast, frequency intakes of these food groups and dietary practices from the VCSDQ were all less than once a day. Conversely, the daily intakes of fruits, vegetables, and water estimated from the VCSDQ were significantly higher than from the 24 h recalls.

Table 3. Median frequency intake and frequency of dietary practices per day from the 1st VCSDQ and 24 h recalls (*n* = 144).

Item	1st VCSDQ			24 h Recalls × 3 Days			*p*-Value *
	Median	25th	75th	Median	25th	75th	
Core food groups (times/day)							
Eating grains	0.64	0.28	2.5	3.00	2.64	3.61	<0.001
Eating vegetables	0.61	0.14	1.00	2.00	1.46	2.36	<0.001
Fruits	0.86	0.43	1.69	1.00	0.64	1.36	0.198
Meat and alternatives	0.83	0.43	1.76	3.00	2.61	3.48	<0.001
Dairy	0.79	0.14	1.22	1.07	0.64	1.71	0.002
Non-core food groups (times/day)							
Drinking sweetened beverages	0.14	0.07	0.43	1.00	0.54	1.71	<0.001
Snacks	0.28	0.14	0.86	1.43	1.00	2.00	<0.001
Eating fast food	0.07	0.00	0.14	0.36	0.00	0.64	<0.001
Eating instant noodles	0.14	0.07	0.43	0.00	0.00	0.5	0.183
Eating processed meat	0.14	0.00	0.43	0.36	0.29	0.71	<0.001
Dietary practices							
Adding salt	0.57	0.28	1.14	0.29	0.00	0.71	<0.001
Watching screens while eating	0.14	0.00	0.79	0.36	0.00	1.07	<0.001
Eating with other family members	0.79	0.14	1.00	1.29	1.00	1.82	<0.001
Eating outside of the home	0.14	0.14	0.43	1.36	1.00	2.00	<0.001
Skipping meals	0.00	0.00	0.14	0.00	0.00	0.29	1.000
Daily intakes							
Fruits (serves/day)	2.00	1.00	2.00	0.73	0.36	1.17	<0.001
Vegetables (serves/day)	2.00	1.00	3.00	0.98	0.69	1.37	<0.001
Water (cups/day)	4.50	2.50	6.50	2.84	1.96	3.71	<0.001

* Wilcoxon matched-pairs signed-rank test *p* < 0.05 significant, one serve = 100 g, one cup = 200 mL.

3.3. Reproducibility

The VCSDQ had moderate to good reproducibility with an ICC of 12 out of 15 items >0.5 (Table 4). The frequency of watching television or other screens while eating had the highest reproducibility value with an ICC = 0.84 (0.78–0.89). Three items with poor reproducibility were intake frequencies for grains, processed meats, and the frequency of eating outside of the home (ICC < 0.5).

Table 4. Reproducibility of the VCSDQ (n = 139).

Item	SCC (95%CI)	ICC (95%CI)	Weighted κ (95%CI)	Exact Agreement (%)	Exact and Adjacent Agreement (%)	GM (%)
Core food groups						
Eating grains	0.20 (0.02–0.38)	0.33 (0.06–0.52)	0.20 (0.02–0.37)	38	73	11
Eating vegetables	0.36 (0.20–0.52)	0.52 (0.33–0.66)	0.35 (0.19–0.51)	43	77	8
Fruits	0.50 (0.37–0.64)	0.67 (0.53–0.76)	0.50 (0.37–0.63)	43	82	3
Meat and alternatives	0.35 (0.21–0.50)	0.52 (0.33–0.66)	0.35 (0.20–0.51)	37	76	6
Dairy	0.34 (0.18–0.50)	0.51(0.31–0.65)	0.34 (0.18–0.5)	37	78	6
Non-core food groups						
Drinking sweetened beverages	0.36 (0.21–0.51)	0.52 (0.33–0.66)	0.35 (0.20–0.50)	42	72	5
Snacks	0.39 (0.24–0.55)	0.57 (0.40–0.69)	0.39 (0.24–0.55)	39	81	9
Eating fast food	0.44 (0.29–0.58)	0.61 (0.46–0.72)	0.44 (0.29–0.59)	45	80	4
Eating instant noodles	0.42 (0.28 0.57)	0.59 (0.43–0.71)	0.42 (0.27–0.57)	40	83	5
Eating processed meat	0.30 (0.14–0.47)	0.43 (0.21–0.60)	0.27 (0.11–0.44)	35	76	7
Dietary practices						
Adding salt	0.45 (0.31–0.60)	0.63 (0.48–0.73)	0.45 (0.31–0.60)	41	81	4
Watching screens while eating	0.72 (0.63–0.81)	0.84 (0.78–0.89)	0.73 (0.64–0.82)	62	87	1
Having family meal	0.49 (0.34–0.64)	0.64 (0.49–0.74)	0.46 (0.31–0.62)	55	83	6
Eating outside of the home	0.31 (0.15–0.47)	0.47 (0.26–0.62)	0.30 (0.13–0.46)	53	65	14
Skipping meal	0.46 (0.29–0.63)	0.63 (0.48–0.74)	0.45 (0.28–0.63)	72	77	12

SCC: Spearman correlation coefficient, ICC: intra-class correlation coefficient, GM: Gross misclassification.

At an individual level, all items except for eating grains had an acceptable Spearman correlation between the first and second administration (SCC $\geq$ 0.3). Only two items (intake frequency of fruits and frequency of watching television or other screens while eating) had a good correlation (SCC $\geq$ 0.5). In addition, 12 out of 15 items (80%) had a gross misclassification lower than 10%. Further examination of the agreement between the first and second administration of the VCSDQ, the weighted kappa showed that all items had at least fair agreement, one out of four core food groups, three out of five non-core food groups, and four out of five dietary practices had a moderate to a substantial agreement without a chance.

3.4. Relative Validity

Overall, the VCSDQ had a low relative validity in most food group intakes and dietary practices. The de-attenuation Spearman correlation ranged from −0.06 for frequency intake of grains to 0.73 for frequency of watching while eating. Seven out of 15 items had an acceptable association (de-SCC > 0.5) (Table 5). By using weighted kappa, only two items (frequency intake of instant noodles and frequency of watching while eating) had a moderate to good relative validity (weighted kappa = 0.43 (0.29–0.57), 0.69 (0.59–0.8), respectively). Although having a fair agreement (0.2 < weighted kappa < 0.4) in the frequency intake of vegetables, dairy, sweetened beverages, processed meat (weighed kappa = 0.23, 0.3, 0.29, and 0.21, respectively), the percentages of gross misclassification among these food groups were less than 10% (4%, 7%, 5% and 5%, respectively). The percentage of exact and adjacent agreement ranged from 58–76% in core-food groups, 68–76% in non-core food groups, and 54–90% in dietary practices. Six items with a high percentage of gross misclassification ($\geq$10%) were grains, sweets and savory snacks, adding salt/sauces, having a meal with family, eating outside of the home, and skipping a meal.

Table 5. Relative validity of the VCSDQ against 24 h recalls (*n* = 144).

Item	SCC	De-SCC *	ICC	ICC-Adjusted *	Weighted κ (95%CI)	EG (%)	EAG (%)	GM (%)
Core food groups								
Eating grains	−0.06 (−0.23–0.10)	−0.06	−0.19 (−0.65–0.14)	−0.20	−0.09 (−0.25–0.07)	23	58	15
Eating vegetables	0.23 (0.08–0.38)	0.25	0.38 (0.14–0.55)	0.41	0.23 (0.08–0.38)	31	70	4
Fruits	0.17 (0.02–0.33)	0.19	0.30 (0.02–0.49)	0.33	0.17 (0.02–0.33)	26	71	9
Meat and alternatives	0.16 (−1.50–0.32)	0.18	0.28 (−0.01–0.48)	0.31	0.16 (0.00–0.31)	26	65	7
Dairy	0.31 (0.15–0.46)	0.33	0.46 (0.26–0.61)	0.49	0.30 (0.15–0.46)	33	76	7
Non-core food groups								
Drinking sweetened beverages	0.30 (0.15–0.45)	0.34	0.45 (0.24–0.60)	0.52	0.29 (0.14–0.43)	33	68	5
Snacks	0.24 (0.08–0.40)	0.25	0.39 (0.15–0.56)	0.41	0.24 (0.09–0.40)	31	68	10
Eating fast food	0.13 (−0.02–0.29)	0.14	0.25 (−0.04–0.46)	0.27	0.14 (−0.01–0.30)	24	69	9
Eating instant noodles	0.44 (0.30–0.57)	0.49	0.60 (0.44–0.71)	0.67	0.43 (0.29–0.57)	43	76	6
Eating processed meat	0.20 (0.06–0.35)	0.22	0.35 (0.10–0.53)	0.38	0.21 (0.05–0.37)	33	75	5
Dietary practices								
Adding salt	0.09 (−0.08–0.25)	0.10	0.18 (−0.14–0.41)	0.20	0.10 (−0.07–0.27)	29	67	13
Watching screens while eating	0.71 (0.61–0.80)	0.73	0.82 (0.75–0.87)	0.84	0.69 (0.59–0.80)	56	90	3
Having family meal	0.17 (0.00–0.33)	0.18	0.25 (−0.05–0.46)	0.27	0.14 (−0.02–0.30)	31	65	12
Eating outside of the home	0.03 (−0.13–0.20)	0.03	0.06 (−0.30–0.33)	0.06	0.03 (−0.13–0.19)	32	54	19
Skipping meal	0.06 (−0.11–0.23)	0.07	0.12 (−0.22–0.37)	0.14	0.08 (−0.09–0.26)	56	60	23

* The correlation was adjusted by within-variation of 3 days 24 h recall; SCC: Spearman rank correlation coefficient; ICC: intra-class correlation; De-SCC: de-attenuation; EC: exact agreement; EAG: exact and adjacent agreement; GM: gross misclassification.

At the group-level, only median of frequency intake of fruits and the frequency of skipping meals were equal (Table 3). In addition, the examination of agreement in the frequency of intakes and dietary practices between the VCSDQ and 24 h recalls using ICC indicated that only three items (intake frequency of sweetened beverages, instant noodles, and frequency of watching television and other screens while eating) had a moderate to a good agreement (Table 5).

The Bland–Altman plots with mean differences in times/day (VCSDQ minus 24 h recalls) and 95% limit of agreements are presented in Figure 3. The VCSDQ underreported frequency intake of eight out of ten food groups and three out of five dietary practices, and overreported frequency intake of fruits, instant noodles, and skipping meals. The mean differences were large for frequency intake of grains (−1.82 times/day) and meat and alternatives (−1.78 times/day) and were small for frequency of skipping meals (0.03 times/day); instant noodles (0.11 times/day) and processed meat (−0.11 times/day). Five items (frequency intake of fast food, instant noodles, processed meat; and frequency of watching while eating and skipping meal) had a 95% limit of agreement within 2 times/day.

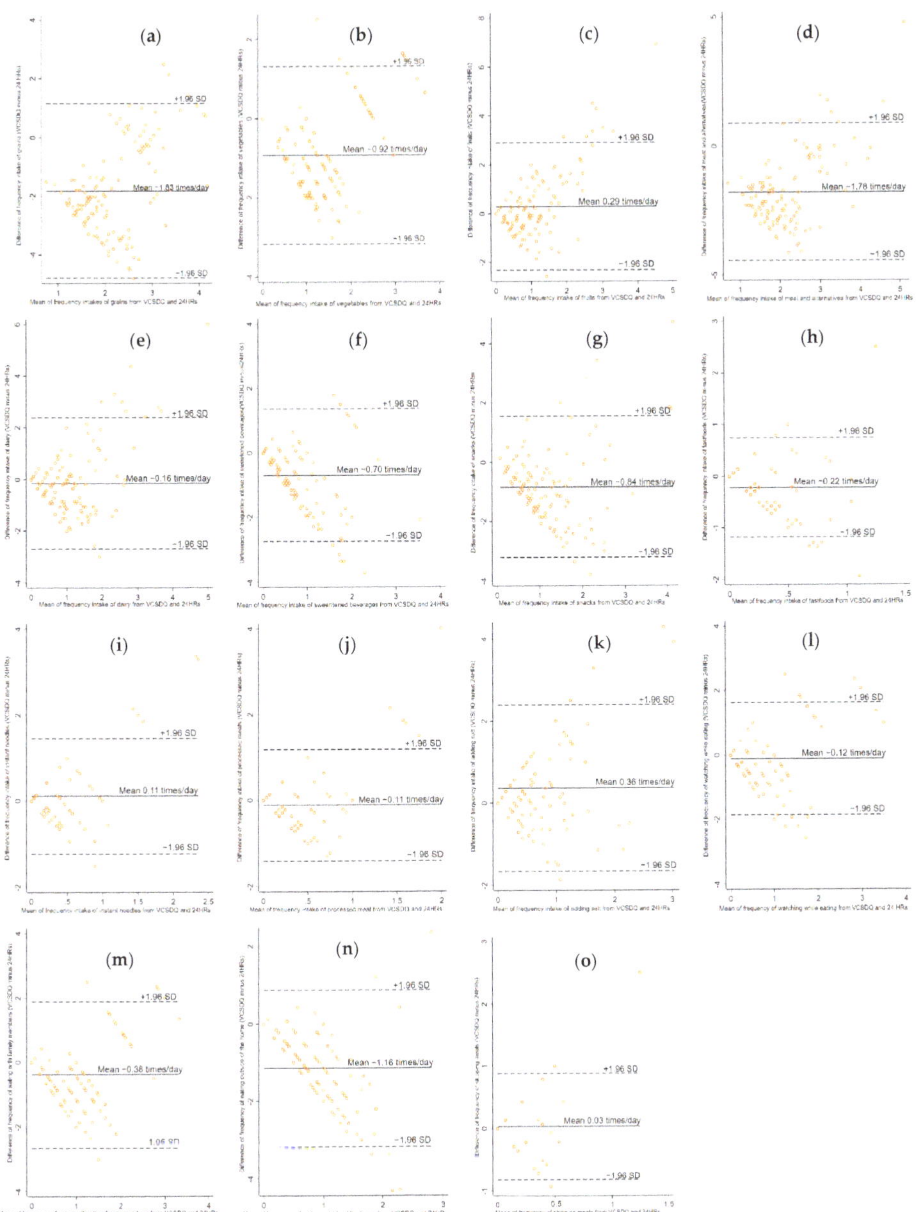

Figure 3. Bland–Altman plots describing the mean differences (VCSDQ minus 24 h recalls) for frequency intake of food groups and dietary practices in times per day: (**a**) grains; (**b**) vegetables; (**c**) fruits; (**d**) meat and alternatives; (**e**) dairy; (**f**) sweetened beverages; (**g**) snacks and discretionary foods; (**h**) fastfoods; (**i**) instant noodles; (**j**) processed meats; (**k**) adding salt; (**l**) watching screens while eating; (**m**) eating with family members; (**n**) eating outside of the home; (**o**) skipping meals. The solid line represents the mean, and the dashed lines represent the 95% limit of agreement (+1.96 SD and −1.96 SD) of the observation. The *y*-axis shows the VCSDQ (test method) minus 24 h recalls (reference method) in times/day, the *x*-axis shows the mean between VCSDQ and 24 h recalls in times/day.

3.5. Direct Validity

The ability to detect the trend in frequency intakes of core and non-core food groups is presented in Table 6. Overall, there was an increase in the amount of food intake (g/mL per day) from 24 h recalls between the first quartile and fourth quartile in six out of 10 food groups (vegetables, fruits, dairy, sweetened beverages, instant noodles, processed meat). For example, the amount of intake of sweetened beverages increased by each quartile from the first quartile (178 mL/day) to the second quartile (257 mL/day) to the third quartile (264 mL/day) and to fourth quartile (333 mL/day) indicating a positive trend between the amount intake of sweetened beverages from 24 h-recalls and the frequency intake from the VCSDQ.

Table 6. Daily food group intakes from 24 h-recalls and by quartiles from the 1st VCSDQ.

Item	Weight (g/Day)			Mean Intakes (g/Day) from 24 h Recall by Quartiles of the 1st SDQ				*p*-Value *
	Mean	25th	75th	Q1	Q2	Q3	Q4	
Core food groups								
Eating grains	212.17	157.75	254.33	213.71	220.50	210.61	206.10	0.650
Eating vegetables	98.15	68.80	136.61	88.19	93.14	115.83	138.31	<0.001
Fruits	73.45	36.01	117.90	69.02	81.39	90.70	115.80	<0.001
Meat and alternatives	215.42	169.98	268.67	216.38	223.97	225.40	231.16	0.920
Dairy	166.07	78.57	269.46	107.76	192.21	203.44	246.11	<0.001
Non-core food groups								
Drinking sweetened beverages	196.20	72.50	368.75	177.98	256.86	263.55	332.99	<0.001
Snacks	67.23	31.96	97.53	60.75	68.59	77.05	76.22	0.160
Eating fast food	22.26	0.00	45.52	24.99	39.28	40.23	34.77	0.070
Eating instant noodles	0.00	0.00	30.38	8.41	17.41	25.71	29.29	<0.001
Eating processed meat	7.20	0.12	14.05	5.94	8.76	9.29	11.93	0.020

* Wilcoxon test for trend using command "nptrend" from Stata17.

The direct validity was also examined with the daily intake of fruits, vegetables, and water. Table 7 shows an increase in the daily amount of intake of fruits and vegetables from the 24 h-recalls corresponding to the increase in serves of fruits and vegetable intakes from the VCSDQ ($p < 0.05$). However, the weight of daily intakes from the 24 h recalls for each response category did not precisely match the servings' illustrations from the VCSDQ. For example, the average intakes of fruits and vegetables from the VCSDQ in the category 4 serves/day or more were defined as 400 g/day or more, whereas these figures from the 24 h recalls were 82.4 g and 145 g, respectively. Although there was a lower water intake in the category "not drinking water" compared to other categories, this item did not show trend ($p = 0.163$). Noticeably, the amount of water intake in categories ">8 cups/day" from the 24 h recall was about 686 mL/day, which is far lower than the estimated water intake for this category (1600 mL/day).

A summary table of results is presented in Table 8 to facilitate the utilization of each questionnaire item.

Table 7. The correlation between daily intakes of fruits, vegetables, and water from the 24 h recalls with the number of serves of fruits/vegetables and cups of water from the VCSDQ (n = 144).

	Responses from VCSDQ		1	2	3	4	5	6
			Not Eating	Less Than 1 Serve	1 Serve	2 Serves	3 Serves	≥4 Serves
Daily intakes from 24 h-recalls × 3 days	Fruits	n	7	15	43	46	23	10
		Mean (g/day)	29.5	69.7	87.7	98.6	103.1	82.4
		p-value *			0.032			
	Vegetables	n	10	19	37	36	20	22
		Mean (g/day)	61.1	84.2	99.1	115.4	115.1	145.0
		p-value *			<0.001			
			Not Drinking	≤1 Cup	2–3 Cups	4–5 Cups	6–7 Cups	≥8 Cups
	Water	n	2	19	40	40	24	15
		Mean (mL/day)	276.8	569.5	584.7	550.3	615.5	606.6
		p-value *			0.163			

* Wilcoxon test for trend using command "nptrend" from Stata 17.

Table 8. Reproducibility, relative validity and sensitive to trend of each VCSDQ items.

	Reproducibility						Validity				Sensitivity to Trend	Usefulness
	Group Level	Individual Level					Group Level	Individual Level				
	ICC	SCC	Cross-Classification		Weighted Kappa	Signed Test	De-SCC	Cross-Classification		Weighted Kappa		
			GM < 10%	EG > 50%				GM < 10%	EG > 50%			
Core food groups												
Eating grains		*			*							NA
Eating vegetables	*	*	*		*		*	*		*	*	RVS
Fruits	*	**	*		**	*		*			*	RVS
Meat and alternatives	*	*	*		*			*				R
Dairy	*	*	*		*		*	*		*	*	RVS
Non-core food groups												
Drinking sweetened beverages	*	*	*		*		*	*		*	*	RVS
Snacks	*	*	*		*		*			*		RV
Eating fast food	*	*	*		**			*				R
Eating instant noodles	*	*	*		**	*	*	*		**	*	RVS
Eating processed meat		*	*		*		*	*		*	*	VS
Dietary practices												
Adding salt	*	*	*		**							R
Watching screens while eating	**	**	*	*	***		**	*		***		RV
Eating with other family members	**	*	*	*	**							R
Eating outside of the home		*		*	*	*				*		V
Skipping meals	*	*		*	**							R
Daily intakes												
Fruits (serves/day)											*	S
Vegetables (serves/day)											*	S
Water (cups/day)												NA

* ICC 0.50–0.75 (moderate intra-class correlation), SCC or de-SCC 0.20–0.49 (acceptable outcomes), cross-classification (good agreement), weighted kappa 0.21–0.4 (fair agreement), signed-test (p > 0.05). ** ICC 0.75–0.90 (good intra-class correlation), SCC or de-SCC > 0.5 (good outcome), weighted kappa 0.41–0.60 (moderate agreement). *** weighted kappa 0.61–0.80 (substantial agreement). NA: non-acceptable for use, R: acceptable reproducibility at group level, V: acceptable validity at group level or individual level, S: acceptable sensitiveness to trend.

4. Discussion

This study examines the reproducibility and relative validity of a 26-item Vietnamese Children's Short Dietary Questionnaire (VCSDQ), which was developed to rapidly evaluate the frequency intakes of five core and five non-core food groups and five dietary practices among children 9–11 years in urban Vietnam over a one-week period.

4.1. Reproducibility

Overall, the VCSDQ showed moderate to good agreement for repeated measurement at a group level (ICC > 0.5) for most of the items (except for grains, processed meat and eating outside of the home) and fair to good agreement for repeated measurement at the individual level (weighted kappa > 0.2). These findings are comparable with another short dietary questionnaire in children and adolescents [9]. The low reproducibility of VCSDQ for processed meat (ICC = 0.43) was similar to the results reported in a questionnaire developed for children 9–10 years old in New Zealand (NZ) (ICC = 0.38) [26]. For the intake of grains, there were no other published studies that we could compare our results with. The most similar study was conducted in NZ children showed that the reproducibility of intake estimates for rice and rice-based dishes was high (ICC = 0.73), whereas in our study, the reproducibility for intakes of grains including rice, rice-based dishes, bread, starchy foods was low (ICC = 0.33). This difference in observations between those two countries may be due to the differences in children's diets between NZ and Vietnam. In NZ rice and rice-based dishes are potentially requested dishes eaten in response to children's preferences so they may to more likely to be remembered [41]. Conversely in Vietnam, rice and rice-based dishes are staple foods and consumed regularly in the diet. Using the weighted kappa, the reproducibility of cereal/grain intakes (k = 0.2) in our study was lower than that observed in older children (11–12 year old and 13–14 year old) in Belgium (k = 0.55 and 0.58) [42] and (12–17 year old) in China (ICC = 0.48) [43]. This higher reproducibility with older age groups could imply that the cognitive capacity of children aged 9–10 may not be fully adequate to quantify the frequency of grain/cereal intake. In addition, the fact that grains (rice, rice noodles, rice paper rolls, etc.) are usually include a range of food items (which may be difficult for young children to discern), are often eaten in different eating contexts and in mixed dishes [9]. This makes the quantification of this food group more difficult, particularly for children. Low reproducibility of estimating cereal intake was also reported in pregnant women (ICC = 0.25) and lactating women in China (ICC = 0.48) [44,45]. Thus, it is possible that estimating the intake of grains and cereals has low reproducibility in Asian countries.

For the reproducibility of dietary practices, four of five items had good reproducibility except for eating outside of the home (ICC = 0.47, k = 0.3). Presently there are no other studies to compare these results with directly. For a similar concept of eating "take away foods", the reproducibility was low in Australian Aboriginal children (k = 0.39) but higher in non-Aboriginal children (k = 0.59) [40]. In China, eating outside of the home was positively associated with overweight and obesity in children (6–17 years old) [40], and with dietary energy from fat and high sugar intake in Vietnamese adolescents (15–17 years old) [46]. However, the questions for eating outside of the home had not been validated or had been drawn from 24 h recalls [46,47]. Thus, apart from examining food intake, questionnaires to examine dietary practices potentially highly relevant to children's nutritional status and health should be validated to enable high-quality dietary data collection in the context of the nutrition transition in low- and middle-income countries.

4.2. Relative Validity

Overall, the relative validity of this first version of the VCSDQ was generally low, with eight out 15 items having a poor agreement, five having a fair agreement, and two having a moderate to good agreement at the individual level. Three out of 15 items had a good agreement at the group level. Similarly, low validity of short dietary questionnaire items was found in other studies [9,26].

For the agreement at the individual level, two core food groups (frequency intakes of vegetables, dairy) and three non-core food groups (sweetened beverages, snacks, processed meat) had a fair agreement (k = 0.21–0.4), one non-core food group (instant noodles) had a moderate agreement (k = 0.43), and one dietary practice (watching screens while eating) had a good agreement (k = 0.69). Although frequency intakes of four non-core food groups and frequency of watching screens while eating were lower than frequency intakes of core-food groups, more items in non-core food groups have a higher validity than core-food groups. This comparison could indicate that these items are children's food preferences, so they are more likely to report them accurately than core-food groups [48]. In the context of high prevalence of overweight and obesity in children, the potential preference for non-core foods as well as the inaccuracy in estimating core-food groups are both issues of concern impacting on reporting. Although children's nutritional status can influence children's reporting of dietary intake [41], this analysis has not yet been completed for this study. Further analysis needs to be conducted to examine factors associated with the accuracy of child report of core and non-core foods groups in the context of high prevalence of overweight and obesity.

Eight out of ten food groups (vegetables, fruits, meat and alternatives, dairy, sweetened beverages, fast-foods, instant noodles, processed meat) and one out of five dietary practices (watching screens while eating) had gross misclassification of less than 10%. All items were able to allocate more than 50% of individuals in the same or adjacent group, which indicated acceptable agreement between the two methods at the individual level. So, although having a fair agreement (weighted kappa = 0.21–0.4), above nine items with misclassification less than 10% or exact/adjacent agreement more than 50% could be fairly used to classify children's intakes from the VCSDQ in a similar group (exact or one quartile difference) from the 24 h recalls. Compared with other studies, our study had a lower gross misclassification than two studies from New Zealand and Belgium [26] and had similar results to a study in China [43]. Although the Spearman correlation coefficient is not a standard statistic for evaluating the agreement between two methods, we used this to compare with other similar studies. Our study had comparable results for most food groups, including fruits, vegetables, meats and alternatives, sweetened beverages, snacks, and adding sauces but lower values for processed meat, rice and rice-based dishes, dairy, and higher values for instant noodles compared to a study in New Zealand [26].

For agreement at the group level, three items (frequency of intake of fruits, instant noodles, and eating outside of home) had good agreement, whereas the median intake of most food groups and dietary practices from the VCSDQ were significantly lower than from the 24 h recalls. The Bland–Altman plots also indicated that the VCSDQ underreported the frequency of food group intakes and dietary practices compared to 24 h recalls.

The explanation for the good agreement of these items could again be due to children's food preferences and the regular consumption of fruits and instant noodles [41,48]. Although the frequency of fruits and vegetables intakes was not high, these food items may be eaten on a regular basis at similar times, for example fruits were often eaten after meals as desert and instant noodles were often eaten in the break time at school or at breakfast/supper. In addition, if children have a specific preference for fruits and instant noodles (either liking or disliking), it is potentially easier to recall the frequency intakes of these food items [41]. Reporting fruit intake is also more accurate among children who are overweight or obese [49], so this could be one of explanation in our study sample given that about 60% of participants were either overweight or obesity. In Ho Chi Minh City eating outside of the home is common dietary practice, which is associated with special events in the family or associated with children's requests restaurants/shops due to the foods or location or environments of eating [50]. So, this dietary practice is potentially more likely to be accurately reported by children.

For the underestimation of frequency intakes from the VCSDQ, one of potential explanation is the retention interval [41]. Children may be more accurately remember what they ate during 24 h recall than over one-week recall from the VCSDQ. In addition, the

24 h recall was administrated by interviewer with the prompt and aids to recall, whereas the VCSDQ was self-administered by children. As a result, most of food items or dietary practices would be omitted in the VCSDQ. In addition, children also had a lower capacity to remember food items when eating out of the home [51], so the food items associated with dietary practices may be also omitted in the VCSDQ. Another possible explanation could also be due to the high prevalence of overweight and obesity as those children are likely to underreport their diet [52].

Although the food frequency questionnaire tends to overestimate children's intake, the under or overreporting of frequency intake of food group and dietary practices from the short dietary questions has not been fully investigated. However, one study in Australian children (4–11 years) reported by parents found that the amount of food group intakes (fruits, vegetables, bread and cereals, meat and alternatives, dairy and extra foods) and diet index score from short food questions were significantly higher than from 24 h recalls [53]. The overreporting of daily amount of intake was similar our study where the daily intake of fruits, vegetables, and water from the VCSDQ were significantly higher than from the 24 h recalls. However, in our study, the frequency intake of food groups from VCSDQ was significantly higher than from 24 h recalls and the underreport tends to be larger in main food groups including grains, meat and alternatives. So, future analysis should examine factors associated with misreporting from short dietary questions.

The frequency of grains, meat and alternatives intakes had a low validity at both the group and individual levels. In other studies, the validity of meat and grain intakes was also low. The Pearson correlation coefficients between a short food questionnaire and three 24 h recalls in Australian children (4–11 years old) assessed as servings/day were 0.08 and 0.07, respectively [53]. The same comparisons in Chinese children (12–17 years old) were 0.13 (cereals), 0.37 (red meat), −0.04 (poultry), 0.14 (seafoods) [43]. A possible explanation for the varying levels of agreement for each questionnaire item is the number of items belonging to a food group and the culture of shared meals in Asian countries. Items with many sub-groups and a wider variety of foods such as grains or meats and meat alternatives seem to have lower validity, whereas items that are clearly defined such as instant noodles had a higher validity. In addition, grains (particularly rice and rice-based dishes) and meat and alternatives (main courses) were often eaten in a shared meal with other family members. So, children may find it more difficult to recall accurately which types of foods and how much of these foods they consumed.

Consequently, to improve the validity of the VCSDQ, food groups such as grains and meat and alternatives, a strategy could be to divide them into discrete sub-groups that more clearly indicate the most obvious and commonly consumed food items within that group using a standard recipe and serving. Additional cognitive interviewing is also recommended to understand children's perceptions of the classification of foods and their portion sizes. In addition, if the data is available, a food list should ideally be developed from the actual dietary intakes of children to provide examples of common dishes and food items. Such a food list should preferably be compiled from data collected in the sub-group of the total population of interest before developing the short dietary questions for dietary assessment [54,55]. These questionnaire modifications could improve the validity of items currently showing low validity in the VCSDQ.

Validity of frequency estimates of dietary practices such as adding salt/sauces, eating with family members, and skipping meals is low. So far, these dietary practices have not been items of interest in other validation studies. In the New Zealand study discussed above, the questionnaire item "frequency intakes of tomato sauces, ketchup" had a low validity similar to our sauces-use question (SCC = −0.11) [26]. In Vietnam, these sauces are often eaten with fried foods (fried meatballs, fried chicken, chips) or fast foods (pizza, burgers). Apart from tomato sauces, fish sauces or soy sauces are often presented on the table for dipping with foods (fried foods or steamed foods) or adding to a dish (pho, rice noodles) as a typical eating habit in Vietnamese meals. In the validation of dietary practices, the 'watching screens while eating' item had high validity, whereas other dietary practices associated with mealtimes had low validity. So, it is likely that if children are not conscious about their diet or are watching screens while eating, they may not be conscious of whether they add sauces to their food or whom they eat with, thereby influencing their answers in the VCSDQ. In addition, eating with family and skipping meals are two behaviors associated with mealtimes, so it is possible that the perception of mealtimes among children in this age group is not clearly defined. In the context of dynamic family activities and work patterns in urban areas, the structure of mealtimes may not be clear and inconsistent day-by-day, so this may also make it more difficult for children to define mealtime practices. This lack of consistency in the diet can also lead to lower accuracy in estimating food intakes and dietary practices [48].

4.3. Direct Validity

Although there are some limitations in the relative validity, the VCSDQ did have the ability to detect trends in the frequency of intakes of dairy, sweetened beverages, instant noodles, and processed meats from low to high quartiles, and for both frequency of intakes and mean daily intakes of fruits and vegetables. These results are comparable to a study among children in Australia (10–12 years old), where mean food intakes from 24 h recalls increased with increasing intake frequencies of pasta/rice, fruit, milk, cheese, butter, red meat, eggs, fruit juice, soft drink, salty snacks, confectionery, and breakfast cereal from the short food frequency questionnaire [40]. So, these items could potentially be used to discriminate the food intake levels of children between low and high in the context of the nutrition transition, where monitoring trends in children's food intakes over time is essential. Such information is needed to develop timely policies and interventions for promoting healthy core food group consumption (fruits, vegetables, dairy), and reduce the intake of unhealthy food groups (e.g., sweetened beverages, instant noodles, and processed meat).

4.4. Strengths and Limitations

4.4.1. Strengths

To date, this is the first study in Vietnam and one of few studies in low- and middle-income countries to develop and validate short dietary questions in evaluating children's whole diet, including food group intakes and dietary practices in school-aged children aged 9–11 years old.

This study included a sample size comparable to other validation studies (*n* = 144), and this sample was representative of fifth-grade students in Ho Chi Minh City in terms of age, sex, and nutritional status [4,15].

Similar to other validation studies of short dietary questions, this study has used multiple analyses (ICC, SCC, weighed kappa, cross-classification, Bland–Altman plots, Wilcoxon test for trend) to examine the reproducibility and validity of the short dietary questions, thereby providing a comprehensive assessment of this tool.

4.4.2. Limitations

This study's data collection method highly depended on children's recall for both the reference method (24 h recall) and the test method (short dietary questionnaire). For the 24 h recall, children were interviewed by trained research assistants using a standard protocol commonly used for 24 h recall interviewing, and the record of 24 h recall was checked by health officers experienced with dietary data collection however, any recall bias due to children's capacity to remember and retrieve their diet and dietary habits could not be eliminated.

Another limitation of this study is the cognitive capacity of children 9–11 years old to report their own diet. Although the accuracy to recall children's diet increase by age [41], generally, the ability to report diet among children under twelve-year-old is potentially limited, particularly in estimating portion size [56]. In addition, although children 8–11 years old could report their own diet using food frequency questionnaire better than their parents [16], they still have limited capacity to report their diet particularly if they ate meals outside of the home [51].

Evidence indicates that the use of photos of foods in dietary intake tools can improve the capacity to correctly estimate food portion size when children report their usual diet [57]. We were able to use a number of photographs of typical food and portion sizes, but due to the continuously increasing diversity of food items available, pictures of food portion sizes were not available for all foods to help students correctly estimate the amount of food typically consumed. The available Vietnamese food composition table lacked many of these newly available food items (such as breakfast cereal, hamburger, pizza, meatballs, etc.), so the classification of these food items into food groups from 24 h recall may mismatch with children's classification of food from the VCSDQ. Such food classification errors are common in low-middle income countries where food composition databases are often insufficient [58].

The cognitive interviewing in this study was conducted with three children who lived in urban areas which may not allow comparison to the perceptions of children living in rural districts where accessibility to fast-food chain restaurants or food items from stores is limited. Therefore, children may have overlooked and misreported their intakes and dietary practices, despite the initial training using example questions. Future development of dietary tools for use in children from low- and middle-income countries should consider cognitive interviewing in a larger sample of children to understand more fully perceptions of food group classifications and portion size estimations as a process of the validation study.

The typical Vietnamese diet includes many mixed dishes and composite foods, so children may find it challenging to classify food groups contained in these mixed dishes, which may also have led to some of the misreporting in this study. A more dish-based assessment could be considered when revising the VCSDQ items as this approach is increasingly used in Asian countries where mixed-dishes are popular [59].

Due to the COVID-19 pandemic, data collection was constrained to a short period of time, and this may have created some level of pressure on the children for self-completion of VCSDQ. In addition, during this time, all interviewers wore masks to prevent COVID-19 transmission, so the ability to build trust and have a comprehensive conversation with children was limited compared to normal circumstances. These conditions may have influenced data collection in this study though this is estimated to be minimal.

Finally, the second VCSDQ was administrated one week after the first VCSDQ, so children may remember what they reported from the first administration. This may lead to the potential increase in the reproducibility of the VCSDQ. However, the increase in interval would reduce the feasibility of the study due to the limited time of data collection and reduce the reproducibility of the VCSDQ due to the change in children's diet week by week.

5. Conclusions

This VCSDQ is one of the first short tools developed in Vietnam. Elements of this tool could be used to evaluate food group intakes and dietary practices in children 9–11 years old. The 26-item VCSDQ had an acceptable reproducibility for all food groups and dietary practices. At the group level, the VCSDQ could be used to rank the frequency intakes of fruits, vegetables, dairy, sweetened beverages, instant noodles, and processed meat. At an individual level, this tool had a fair to good capacity to evaluate frequency intakes of vegetables, dairy sweetened beverages, snacks, instant noodles, processed meat (and frequency of watching screens while eating). Revisions to the VCSDQ need to be made in order to use it as a questionnaire to evaluate dietary intakes and dietary practices among children. These revisions will provide real time data for the development and evaluation of policies and interventions in low- and middle-income countries.

Supplementary Materials: The following supporting information can be downloaded at: https://www.mdpi.com/article/10.3390/nu14193996/s1, Table S1: STROBE-nut: An extension of the

STROBE statement for nutritional epidemiology; Figure S1: The Vietnamese Children's Short Dietary Questionnaire [60].

Author Contributions: Conceptualization, T.M.T.M., S.N., J.C.V.d.P. and D.G.; methodology, T.M.T.M. and Q.C.T.; validation, T.M.T.M.; formal analysis, T.M.T.M.; investigation, T.M.T.M.; writing—original draft preparation, T.M.T.M.; writing—review and editing T.M.T.M., Q.C.T., S.N., J.C.V.d.P. and D.G. All authors have read and agreed to the published version of the manuscript.

Funding: This research was funded by Australian Government Research Training Program, and QUT HDR Tuition Fee Scholarship to T. M. T. M. for the "Doctor of Philosophy program at the Queensland University of Technology, Brisbane, Australia" and "The APC was funded by the Queensland University of Technology".

Institutional Review Board Statement: The study was conducted in accordance with the Declaration of Helsinki and approved by the Human Research Ethics Committee of Queensland University of Technology, Brisbane, Australia (protocol version 3, approved 30 September 2019; UHREC Reference number: 1900000601).

Informed Consent Statement: Informed consent was obtained from all children and their parents involved in the study.

Data Availability Statement: The datasets generated and analyzed during the current study is available from the corresponding author on reasonable request.

Acknowledgments: We would like to thank all children participating in this research and research team for collecting data. In addition, we also send our special thanks to Ho Chi Minh City Nutrition Center for giving me an opportunity to study PhD, Ho Chi Minh Center for Disease Control and Department of Education for giving us the permission to approach schools for this study. Finally, we are grateful for the contribution from Do Thi Ngoc Diep, Tran Thi Minh Hanh, Vu Quynh Hoa, Tang Kim Hong and Pham Thi Lan Anh in the process of developing the short dietary questionnaire.

Conflicts of Interest: The authors declare no conflict of interest.

References

1. Word Health Organization. Obesity and Overweight. Available online: https://www.who.int/news-room/fact-sheets/detail/obesity-and-overweight (accessed on 4 August 2022).
2. Abarca-Gómez, L.e.a. Worldwide trends in body-mass index, underweight, overweight, and obesity from 1975 to 2016: A pooled analysis of 2416 population-based measurement studies in 128.9 million children, adolescents, and adults. *Lancet* **2017**, *390*, 2627–2642. [CrossRef]
3. Simmonds, M.; Llewellyn, A.; Owen, C.G.; Woolacott, N. Predicting adult obesity from childhood obesity: A systematic review and meta-analysis. *Obes. Rev.* **2016**, *17*, 95–107. [CrossRef] [PubMed]
4. Mai, T.M.T.; Pham, N.O.; Tran, T.M.H.; Baker, P.; Gallegos, D.; Do, T.N.D.; van der Pols, J.C.; Jordan, S.J. The double burden of malnutrition in Vietnamese school-aged children and adolescents: A rapid shift over a decade in Ho Chi Minh City. *Eur. J. Clin. Nutr.* **2020**, *74*, 1448–1456. [CrossRef] [PubMed]
5. Harris, J.; Nguyen, P.H.; Tran, L.M.; Huynh, P.N. Nutrition transition in Vietnam: Changing food supply, food prices, household expenditure, diet and nutrition outcomes. *Food Secur.* **2020**, *12*, 1141–1155. [CrossRef]
6. Song, S.; Song, W.O. National nutrition surveys in Asian countries: Surveillance and monitoring efforts to improve global health. *Asia Pac. J. Clin. Nutr.* **2014**, *23*, 514–523. [CrossRef]
7. Hendrie, G.; Riley, M. Performance of Short Food Questions to Assess Aspects of the Dietary Intake of Australian Children. *Nutrients* **2013**, *5*, 4822–4835. [CrossRef]
8. Flood, V.; Webb, K.; Rangan, A. *Recommendations for Short Questions to Assess Food Consumption in Children for the NSW Health Surveys*; University of Wollongong: Wollongong, Australia, 2005. Available online: https://ro.uow.edu.au/hbspapers/364/ (accessed on 13 September 2019).
9. Golley, R.K.; Bell, L.K.; Hendrie, G.A.; Rangan, A.M.; Spence, A.; McNaughton, S.A.; Carpenter, L.; Allman-Farinelli, M.; Silva, A.; Gill, T.; et al. Validity of short food questionnaire items to measure intake in children and adolescents: A systematic review. *J. Hum. Nutr. Diet* **2017**, *30*, 36–50. [CrossRef]
10. Kirkpatrick, S.I.; Baranowski, T.; Subar, A.F.; Tooze, J.A.; Frongillo, E.A. Best Practices for Conducting and Interpreting Studies to Validate Self-Report Dietary Assessment Methods. *J. Acad. Nutr. Diet.* **2019**, *119*, 1801–1816. [CrossRef]
11. Ho Chi Minh Statistics Office. *Ho Chi Minh City Statistical Yearbook 2017*; Thanh Nien Publishing: Ho Chi Minh City, Vietnam, 2018.
12. General Statistics Office. *Statistical Summary Book of Vietnam 2017*; Statistical Publising House: Ho Chi Minh City, Vietnam, 2017.
13. Cade, J.; Thompson, R.; Burley, V.; Warm, D. Development, validation and utilisation of food-frequency questionnaires—A review. *Public Health Nutr.* **2002**, *5*, 567–587. [CrossRef]

14. Willett, W. *Nutritional Epidemiology*, 3rd ed.; Oxford University Press: New York, NY, USA, 2013.
15. To, Q.G.; Gallegos, D.; Do, D.V.; Tran, H.T.M.; To, K.G.; Wharton, L.; Trost, S.G. The level and pattern of physical activity among fifth-grade students in Ho Chi Minh City, Vietnam. *Public Health* **2018**, *160*, 18–25. [CrossRef]
16. Burrows, T.L.; Collins, C.E.; Truby, H.; Callister, R.; Morgan, P.J.; Davies, P.S.W. Doubly labelled water validation of child versus parent report of total energy intake by food frequency questionnaire. *Australas. Med. J.* **2011**, *4*, 775.
17. National Institute of Nutrition. Vietnamese Food-Based Dietary Guidelines. Available online: http://www.fao.org/3/a-as980o.pdf (accessed on 6 August 2022).
18. National Institute of Nutrtition. Guideline to Use the Food Pyramid for Primary School Children. Available online: http://dinhduonghocduong.net/vi/giao-duc-dinh-duong.nd133/bai-06.i195.html (accessed on 14 April 2019).
19. Polit, D.F.; Beck, C.T.; Owen, S.V. Is the CVI an acceptable indicator of content validity? Appraisal and recommendations. *Res. Nurs. Health* **2007**, *30*, 459–467. [CrossRef] [PubMed]
20. Braintree, M. Key Survey. Available online: https://survey.qut.edu.au/site/ (accessed on 10 September 2019).
21. National Institute of Nutrition. *Book of Picture for Dietary Assessment for Children*; Medical Publishing House: Hanoi, Vietnam, 2014.
22. Ho Chi Minh City Nutrition Center. *Energy and Nutritional Values of Common Street Foods*; Medical Publishing House: Hanoi, Vietnam, 2017.
23. Vu, Q.H.; Do, T.N.D.; Tran, T.M.H.; Phan, T.K.T. Food intake characteristics of children at two primary schools in Ho Chi Minh city. *J. Food Nutr. Sci.* **2014**, *10*, 69–76.
24. National Institute of Nutrition. *Vietnamese Food Composition Table*; Medical Publishing House: Hanoi, Vietnam, 2017.
25. Allan, C.; Abdul Kader, U.H.; Ang, J.Y.Y.; Muhardi, L.; Nambiar, S. Relative validity of a semi-quantitative food frequency questionnaire for Singaporean toddlers aged 15–36 months. *BMC Nutr.* **2018**, *4*, 42. [CrossRef] [PubMed]
26. Saeedi, P.; Skeaff, S.A.; Wong, J.E.; Skidmore, P.M. Reproducibility and Relative Validity of a Short Food Frequency Questionnaire in 9–10 Year-Old Children. *Nutrients* **2016**, *8*, 271. [CrossRef]
27. World Health Organization. *Training Course on Child Growth Assessment*; WHO: Geneva, Switzerland, 2008.
28. de Onis, M.; Onyango, A.W.; Borghi, E.; Siyam, A.; Nishida, C.; Siekmann, J. Development of a WHO growth reference for school-aged children and adolescents. *Bull. World Health Organ.* **2007**, *85*, 660–667. [CrossRef]
29. Goldberg, G.R.; Black, A.E.; Jebb, S.A.; Cole, T.J.; Murgatroyd, P.R.; Coward, W.A.; Prentice, A.M. Critical evaluation of energy intake data using fundamental principles of energy physiology: 1. Derivation of cut-off limits to identify under-recording. *Eur. J. Clin. Nutr.* **1991**, *45*, 569–581.
30. Elliott, S.A.; Davies, P.S.; Nambiar, S.; Truby, H.; Abbott, R.A. A comparison of two screening methods to determine the validity of 24-h food and drink records in children and adolescents. *Eur. J. Clin. Nutr.* **2011**, *65*, 1314–1320. [CrossRef]
31. Maffeis, C.; Schutz, Y.; Micciolo, R.; Zoccante, L.; Pinelli, L. Resting metabolic rate in six- to ten-year-old obese and nonobese children. *J. Pediatr.* **1993**, *122*, 556–562. [CrossRef]
32. Molnár, D.; Jeges, S.; Erhardt, E.; Schutz, Y. Measured and predicted resting metabolic rate in obese and nonobese adolescents. *J. Pediatr.* **1995**, *127*, 571–577. [CrossRef]
33. Black, A.E. Critical evaluation of energy intake using the Goldberg cut-off for energy intake:basal metabolic rate. A practical guide to its calculation, use and limitations. *Int. J. Obes.* **2000**, *24*, 1119–1130. [CrossRef]
34. Koo, T.K.; Li, M.Y. A Guideline of Selecting and Reporting Intraclass Correlation Coefficients for Reliability Research. *J. Chiropr. Med.* **2016**, *15*, 155–163. [CrossRef] [PubMed]
35. Lombard, M.J.; Steyn, N.P.; Charlton, K.E.; Senekal, M. Application and interpretation of multiple statistical tests to evaluate validity of dietary intake assessment methods. *Nutr. J.* **2015**, *14*, 40. [CrossRef] [PubMed]
36. Masson, L.F.; McNeill, G.; Tomany, J.O.; Simpson, J.A.; Peace, H.S.; Wei, L.; Grubb, D.A.; Bolton-Smith, C. Statistical approaches for assessing the relative validity of a food-frequency questionnaire: Use of correlation coefficients and the kappa statistic. *Public Health Nutr.* **2003**, *6*, 313–321. [CrossRef] [PubMed]
37. Gibson, R.S. *Principles of Nutritional Assessment*, 2nd ed.; Oxford University Press: New York, NY, USA, 2005.
38. Bland, J.M.; Altman, D.G. Measuring agreement in method comparison studies. *Stat. Methods Med. Res.* **1999**, *8*, 135–160. [CrossRef]
39. Landis, J.R.; Koch, G.G. The measurement of observer agreement for categorical data. *Biometrics* **1977**, *33*, 159–174. [CrossRef]
40. Gwynn, J.D.; Flood, V.M.; D'Este, C.A.; Attia, J.R.; Turner, N.; Cochrane, J.; Wiggers, J.H. The reliability and validity of a short FFQ among Australian Aboriginal and Torres Strait Islander and non-Indigenous rural children. *Public Health Nutr.* **2010**, *14*, 388–401. [CrossRef]
41. Sharman, S.J.; Skouteris, H.; Powell, M.B.; Watson, B. Factors Related to the Accuracy of Self-Reported Dietary Intake of Children Aged 6 to 12 Years Elicited with Interviews: A Systematic Review. *J. Acad. Nutr. Diet.* **2016**, *116*, 76–114. [CrossRef]
42. Vereecken, C.A.; Maes, L. A Belgian study on the reliability and relative validity of the Health Behaviour in School-Aged Children food-frequency questionnaire. *Public Health Nutr.* **2003**, *6*, 581–588. [CrossRef]
43. Dan, L.I.U.; La Hong, J.U.; Yang, Z.Y.; Zhang, Q.; Gao, J.F.; Di Ping, G.O.N.G.; Guo, D.D.; Luo, S.Q.; Zhao, W.H. Food Frequency Questionnaire for Chinese Children Aged 12–17 Years: Validity and Reliability. *Biomed. Environ. Sci.* **2019**, *32*, 486. [CrossRef]
44. Zhang, H.; Qiu, X.; Zhong, C.; Zhang, K.; Xiao, M.; Yi, N.; Xiong, G.; Wang, J.; Yao, J.; Hao, L.; et al. Reproducibility and relative validity of a semi-quantitative food frequency questionnaire for Chinese pregnant women. *Nutr. J.* **2015**, *14*, 56. [CrossRef] [PubMed]
45. Ding, Y.; Li, F.; Hu, P.; Ye, M.; Xu, F.; Jiang, W.; Yang, Y.; Fu, Y.; Zhu, Y.; Lu, X.; et al. Reproducibility and relative validity of a semi-quantitative food frequency questionnaire for the Chinese lactating mothers. *Nutr. J.* **2021**, *20*, 20. [CrossRef] [PubMed]

46. Lachat, C.; Khanh, L.N.B.; Khan, N.C.; Dung, N.Q.; Van Anh, N.D.; Roberfroid, D.; Kolsteren, P. Eating out of home in Vietnamese adolescents: Socioeconomic factors and dietary associations. *Am. J. Clin. Nutr.* **2009**, *90*, 1648–1655. [CrossRef]
47. Ma, Y.; Gong, W.; Ding, C.; Song, C.; Yuan, F.; Fan, J.; Feng, G.; Chen, Z.; Liu, A. The association between frequency of eating out with overweight and obesity among children aged 6–17 in China: A National Cross-sectional Study. *BMC Public Health* **2021**, *21*, 1005. [CrossRef] [PubMed]
48. Kuzma, J.W.; Lindsted, K.D. Determinants of long-term (24-year) diet recall ability using a 21-item food frequency questionnaire. *Nutr. Cancer* **1989**, *12*, 151–160. [CrossRef]
49. Harrington, K.F.; Kohler, C.L.; McClure, L.A.; Franklin, F.A. Fourth Graders' Reports of Fruit and Vegetable Intake at School Lunch: Does Treatment Assignment Affect Accuracy? *J. Am. Diet. Assoc.* **2009**, *109*, 36–44. [CrossRef]
50. Sahakian, M.; Saloma, C.A.; Erkman, S. Emerging consumerism and eating out in Ho Chi Minh City, Vietnam: The social embeddedness of food sharing. In *Food Consumption in the City: Practices and Patterns in Urban Asia and the Pacific*; Taylor & Francis: Oxfordshire, UK, 2016; Chapter 3.
51. Burrows, T.L.; Martin, R.J.; Collins, C.E. A systematic review of the validity of dietary assessment methods in children when compared with the method of doubly labeled water. *J. Am. Diet Assoc.* **2010**, *110*, 1501–1510. [CrossRef]
52. Collins, C.E.; Watson, J.; Burrows, T. Measuring dietary intake in children and adolescents in the context of overweight and obesity. *Int. J. Obes.* **2010**, *34*, 1103–1115. [CrossRef]
53. Hendrie, G.A.; Viner Smith, E.; Golley, R.K. The reliability and relative validity of a diet index score for 4–11-year-old children derived from a parent-reported short food survey. *Public Health Nutr.* **2014**, *17*, 1486–1497. [CrossRef]
54. Hunsberger, M.; O'Malley, J.; Block, T.; Norris, J.C. Relative validation of Block Kids Food Screener for dietary assessment in children and adolescents. *Matern. Child Nutr.* **2015**, *11*, 260–270. [CrossRef]
55. Mulasi-Pokhriyal, U.; Smith, C. Comparison of the Block Kid's Food Frequency Questionnaire with a 24 h dietary recall methodology among Hmong-American children, 9–18 years of age. *Br. J. Nutr.* **2013**, *109*, 346–352. [CrossRef] [PubMed]
56. Livingstone, M.B.; Robson, P.J.; Wallace, J.M. Issues in dietary intake assessment of children and adolescents. *Br. J. Nutr.* **2004**, *92* (Suppl. 2), S213–S222. [CrossRef] [PubMed]
57. Higgins, J.A.; LaSalle, A.L.; Zhaoxing, P.; Kasten, M.Y.; Bing, K.N.; Ridzon, S.E.; Witten, T.L. Validation of photographic food records in children: Are pictures really worth a thousand words? *Eur. J. Clin. Nutr.* **2009**, *63*, 1025–1033. [CrossRef] [PubMed]
58. Micha, R.; Coates, J.; Leclercq, C.; Charrondiere, U.R.; Mozaffarian, D. Global Dietary Surveillance: Data Gaps and Challenges. *Food Nutr. Bull.* **2018**, *39*, 175–205. [CrossRef]
59. Shinozaki, N.; Yuan, X.; Murakami, K.; Sasaki, S. Development, validation and utilisation of dish-based dietary assessment tools: A scoping review. *Public Health Nutr.* **2021**, *24*, 223–242. [CrossRef] [PubMed]
60. Lachat, C.; Hawwash, D.; Ocké, M.C.; Berg, C.; Forsum, E.; Hörnell, A.; Larsson, C.; Sonestedt, E.; Wirfält, E.; Åkesson, A.; et al. Strengthening the Reporting of Observational Studies in Epidemiology-Nutritional Epidemiology (STROBE-nut): An Extension of the STROBE Statement. *PLoS Med.* **2016**, *13*, e1002036. [CrossRef]

 *nutrients*

Article

Development and Field-Testing of Proposed Food-Based Dietary Guideline Messages and Images amongst Consumers in Tanzania

Lisanne M. Du Plessis [1,*], Nophiwe Job [2], Angela Coetzee [3], Shân Fischer [2], Mercy P. Chikoko [4], Maya Adam [2,5], Penelope Love [6,*] and on behalf of the Food-Based Dietary Guideline (FBDG) Technical Working Group (TWG) Led by Tanzania Food and Nutrition Centre (TFNC) [†]

[1] Division of Human Nutrition, Department of Global Health, Faculty of Medicine and Health Sciences, Stellenbosch University, Cape Town 7500, South Africa
[2] Digital Medic South Africa, Stanford Center for Health Education, Cape Town 8000, South Africa; njob@stanford.edu (N.J.); fischers@stanford.edu (S.F.); madam@stanford.edu (M.A.)
[3] Sustainability Institute, School for Public Leadership, Stellenbosch University, Stellenbosch 7600, South Africa; angela.coetzee@me.com
[4] Food and Agriculture Organisation, Sub-Regional Office for Southern Africa, Harare 3730, Zimbabwe; mercy.chikoko@fao.org
[5] Stanford Center for Health Education, Department of Pediatrics, Stanford School of Medicine, Stanford, CA 94305, USA
[6] Institute for Physical Activity and Nutrition (IPAN), School of Exercise and Nutrition Sciences (SENS), Deakin University, Geelong, VIC 3216, Australia
[*] Correspondence: lmdup@sun.ac.za (L.M.D.P.); penny.love@deakin.edu.au (P.L.); Tel.: +27-21-938-9175 (L.M.D.P.)
[†] Members are listed below in the Acknowledgement.

Citation: Du Plessis, L.M.; Job, N.; Coetzee, A.; Fischer, S.; Chikoko, M.P.; Adam, M.; Love, P.; on behalf of the Food-Based Dietary Guideline (FBDG) Technical Working Group (TWG) Led by Tanzania Food and Nutrition Centre (TFNC). Development and Field-Testing of Proposed Food-Based Dietary Guideline Messages and Images amongst Consumers in Tanzania. *Nutrients* 2022, 14, 2705. https://doi.org/10.3390/nu14132705

Academic Editor: Patrick R.M. Lauwers

Received: 30 April 2022
Accepted: 20 June 2022
Published: 29 June 2022

Publisher's Note: MDPI stays neutral with regard to jurisdictional claims in published maps and institutional affiliations.

Abstract: In this paper we report on the development and field-testing of proposed food-based dietary guideline (FBDG) messages among Tanzanian consumers. The messages were tested for cultural appropriateness, consumer understanding, acceptability, and feasibility. In addition, comprehension of the messages was assessed using culturally representative images for low literacy audiences. Focus group discussions were used as method for data collection. Results indicate that the core meaning of the proposed FBDG messages and images were understood and acceptable to the general population. However, participants felt that nutrition education would be required for improved comprehension. Feasibility was affected by some cultural differences, lack of nutrition knowledge, time constraints, and poverty. Suggestions were made for some rewording of certain messages and editing of certain images. It is recommended that the field-tested messages and images, incorporating the suggested changes, should be adopted. Once adopted, the FBDGs can be used to inform and engage various stakeholders, including parents, caregivers, healthcare providers and educators on appropriate nutritional practices for children and adults. They can also be used to guide implementation of relevant policies and programmes to contribute towards the achievement of sustainable healthy diets and healthy dietary patterns.

Keywords: food-based dietary guidelines; messages; images; pre-testing; Tanzania; nutrition education

1. Introduction

The United Nations Food and Agriculture Organization (FAO) together with the World Health Organization (WHO) developed the concept of food-based dietary guidelines (FBDGs) in 1995 [1]. Country-specific FBDGs are evidence-based recommendations—simple advisory statements, that express the principles of nutrition education mostly in terms of foods, considering customary dietary patterns, ecological setting, socio-economic and cultural factors, and the biological and physical environment in which the population lives. They are intended to guide the broad public to consume a healthy diet which is both health promoting and protective against the development of malnutrition and non-communicable diseases (NCDs) [1,2].

FBDGs include messages and visual illustrations (images) to help consumers to implement them as part of a healthy lifestyle. It translates nutrient standards and recommendations (dietary goals and guidelines) into simple, practical advice on the types, and sometimes quantities, of various foods needed for healthy dietary patterns [3]. The main purpose of the messages and images is therefore to assist the consumer in choosing a diet adequate in nutrients that is protective of malnutrition and NCDs. FBDGs can be accompanied by a food guide, which is a graphic representation or image of some or all FBDGs, either as one single combined illustration or a series of images that indicates the foods and food groups that should be consumed as part of a healthy diet [4]. One way FAO supports member countries is to assist in the development, revision, and implementation of FBDGs and food guides, aligned with the current evidence base. Periodically, FAO conducts reviews on progress in FBDG development and use [1,5].

Close to 100 countries around the globe have developed FBDGs; however, there are currently only nine countries in Africa with finalised dietary guidelines. These countries include Benin, Kenya, Namibia, Nigeria, Seychelles, Sierra Leone, South Africa, Zambia and Ethiopia. With the increasing burden of malnutrition across the African continent, many countries are currently developing their first set of dietary guidelines, and FAO are encouraging more countries to follow suit [2].

In early 2019, the FAO put out a call for proposals to researchers for the development and field-testing of proposed FBDG messages and a food guide for, among others, Tanzania. This project formed part of the FAO's larger objectives to improve diets and nutrition in Southern Africa by developing and testing FBDGs and food guides for various countries and ultimately to contribute to eradicating hunger, malnutrition, and food insecurity.

The Republic of Tanzania is the largest country in East Africa and includes the Mainland and Zanzibar. The country is divided into 21 regions on the Mainland and three on the island (Figure 1). The main local language is Kiswahili. The literacy level is 78%. About 80% of Tanzanians are subsistence farmers or fisherman. Agriculture includes coffee, sisal, tea, cotton, and cattle. The country exports gold, coffee, cashew nuts, manufactured good and cotton [6].

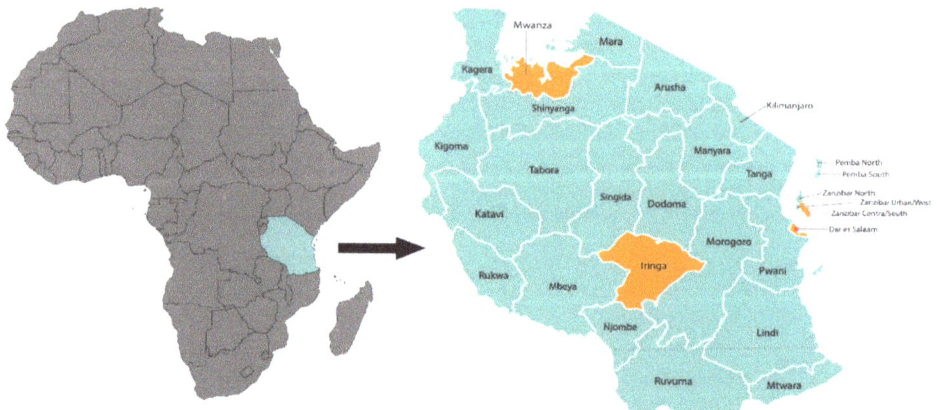

Figure 1. Map of Africa indicating the Republic of Tanzania and country-level districts. Source: https://stock.adobe.com/images (accessed on 26 April 2022).

Since the early 2000s, Tanzania has experienced impressive economic growth, strong resilience to external fiscal shocks and a decline in basic needs poverty because of its vast resources, sociopolitical stability and economic reforms [7–9]. In 2018, the poverty rate declined from 34.4% to 26.4%; however, this reduction was surpassed by the rate of population growth, which resulted in an increase in the absolute number of poor people in Tanzania. Therefore, about 14 million people lived below the national poverty line of 49,320 Tanzanian Shilling per adult equivalent per month and about 26 million (about 49% of the population) lived below the USD 1.90 per person per day international poverty line in 2018 [9]. There are large gaps and disparities in the poverty distribution among geographic regions; however, poverty seems to be most concentrated in the Western and Lake zones of the country and lowest in the Eastern Zones [9]. Promoting healthy diets in low-income and middle-income countries (LMICs) can reduce social inequality in diet between the poor and the rich, especially when it targets disadvantaged population groups and because of both short-term and long-term economic benefits to households due to better health and educational outcomes [10].

Similarly to other LMICs, Tanzania suffers a triple burden of malnutrition (i.e., undernutrition, micronutrient deficiencies and over-nutrition). According to the Tanzania Demographic and Health Survey (TDHS) 2010 [11] and 2015 [12], respectively, while stunting among young children has notably declined from 44% in 2010 to 34.5% in 2015 and further to 31.8% in 2018, figures remain high due to factors such as underweight among mothers, low birth weight, and low dietary diversity [13]. The National Nutrition Study conducted by the Tanzania Food and Nutrition Center in 2018 indicated that 7.3% of women of reproductive age (WRA) were underweight (body mass index (BMI) < 18.5 kg/m^2) and 31.7% overweight (BMI > 25 kg/m^2) [13]. World Bank data from 2019 estimated that NCDs such as cardiovascular diseases, cancers, diabetes, and chronic respiratory diseases may cause 74% of all deaths globally. The figure for Tanzania was 34% [14]. Evidence from the TDHS indicates an increasing prevalence of overweight and obesity; the proportion of women aged 15–49 years who are overweight and obese increased from 22% in 2010 [11] to 28% in 2015 [12] with uneven prevalence distribution between rural and urban areas, and the higher education and wealth quintiles being the most affected. Furthermore, the 2012 Tanzania STEPS study [15] reported the prevalence of hypertension to range from 27.1% to 32.2% and 28.6% to 31.5% in men and women, respectively, and the prevalence of elevated cholesterol to be 26% and obesity to be 35% among adults over 25 years.

An unhealthy diet is one of the most important risk factors that needs to be addressed to tackle the triple burden of malnutrition [16] and diet-related diseases [17,18] in LMICs such as Tanzania, yet there is a scarcity of national data on the patterns of dietary intake. In general, the majority of Tanzanians eat unbalanced diets high in staples (maize and rice) and small amounts of vegetables and fruit. In many areas, people have shifted from traditional diets to consumption of foods high in refined carbohydrates, sugar, fat and salt such as cereals and meat [15,19]. Tanzania's changing food environment is characterized by declining total shares of household income spent on food [20] and increased access to non-staples, processed foods, edible fat and sugary beverages [9,19]. Available information indicates significant dietary gaps, including inadequate intakes of vitamin A, calcium, folate, zinc, iodine, and B vitamins leading to deficiencies, which are of economic and public health concern among all population groups in Tanzania [21]. In sub-Saharan African countries such as Tanzania, micronutrient intake has declined over the past 50 years, as shown by a reduced dietary micronutrient density index (average micronutrient density of the food supply based on 14 micronutrients, calcium, copper, iron, folate, magnesium, niacin, phosphorus, riboflavin, thiamin, vitamin A, vitamin B12, vitamin B6, vitamin C and zinc). Reasons for this include increased availability of grains (rice, maize, and wheat) and vegetable oils which have low micronutrient density, and decreased proportional availability of pulses, dairy products, meat, nuts and seeds, fruit, and vegetables [22]. Recent changes in dietary patterns in LMICs [23,24] as in Tanzania, indicate an important focus on improving diet quality for better health, prevention of diet-related diseases and reduction in the triple burden of malnutrition [24,25].

A critical consideration for nutrition education is that people eat foods and not individual nutrients. FBDGs must therefore be based on (i) the country's nutrition-related public health issues, (ii) the availability, accessibility, and perception of the message content, and (iii) the acceptability to all populations taking into account their lifestyle, cultural eating habits and socio-economic circumstances [1]. Important considerations for Tanzania are the variation in food availability across regions, by season and in rural/urban settings. Another consideration is cultural and religious food preferences and taboos. The Tanzanian cuisine is a fusion of East Asian, Portuguese and Indian cooking and spices. Approximately 61% of the population is Christian, 35% Muslim, and 4% other religious groups. Zanzibar's residents are 99% Muslim [26] and as such, certain food habits are distinct to specific areas of the country. Food systems in Tanzania are also not uniform; they vary according to location, socio-economic status, and food market infrastructure. Various agro-ecological zones have resulted in different dietary patterns. With the rise of supermarkets and fast-food outlets in big cities such as Dar es Salaam and an increasing demand for quality food products in sufficient quantities, a major issue is ensuring that consumers are properly guided on what foods to eat and where to access such foods at reasonable cost [13]. Developing and testing of FBDGs should therefore take cognizance of all these dimensions.

Rationale for this Study

A research team from Stellenbosch University, under the leadership of the principal investigator (PI) and first author from the Division of Human Nutrition, was appointed in 2019 by the FAO Sub-Regional Office for Southern Africa (SFS) to test proposed FBDG messages and images for Tanzania. In this paper we report on the development and field-testing of the proposed FBDGs for cultural appropriateness, consumer understanding, acceptability, and feasibility among Tanzanian consumers. In addition, we assessed the comprehension of accompanying culturally representative visual illustrations (images) for low literacy audiences.

2. Materials and Methods

2.1. Development of the Draft FBDGs for Tanzania

The Tanzanian FBDG Technical Working Group (TWG) comprising of academics and government officials from various sectors, United Nations, and non-governmental organizations from the Republic of Tanzania (Mainland and Zanzibar), led by TFNC in collaboration with FAO, developed a set of 12 FBDG themes (Box 1) in early 2019 informed by public health nutrition and dietary data available for the country. The instruction to the research team was to refine these themes into preliminary FBDG messages. It was also expected that each message would have a visual illustration (image) that would inform one combined illustration (food guide). The research team was requested to support with the pre-testing of the messages and images.

From the 12 themes suggested by FAO and the TWG, the research team proposed preliminary FBDGs. These proposed preliminary FBDGs were based on FAO's characteristics and criteria of FBDG messages [1,2,4], experience with testing FBDG in South Africa [27] and studying relevant scientific literature on the topic as well as country-specific literature for Tanzania.

During meetings in Zanzibar and Dar es Salaam (June 2019), the researchers, FAO delegates and the TGW members workshopped the preliminary FBDG messages. The research team suggested a life cycle approach for the development of images and refinement of the messages. This approach considers messages in all stages of human development, targeting different age groups of the population. The justification for a life cycle approach was based on the substantial, growing evidence of the impact of optimal nutrition during the first 1000 days of life (from conception to two years of age) on the rest of the human life span [28,29].

Box 1. Food-based dietary guideline themes suggested for Tanzania

1. Eating a variety of foods from the following food groups: (a) vegetables, (b) fruits, (c) cereals/grains, starchy tubers or roots, (d) pulses, nuts, and seeds, (e) animal and animal products, milk and milk products, and (f) fruits and vegetables combined;
2. Eating vegetables and fruits. Focusing on green leafy vegetables and orange and yellow coloured, non-citrus fruits and vegetables;
3. Consumption of animal and animal products, milk and milk products;
4. Consumption of pulses, nuts, and seeds daily;
5. Limiting intakes of highly processed foods containing saturated fats and trans-fats, added sugar, and added salt;
6. Infants and young children feeding (exclusive breastfeeding before 6 months and complimentary feeding after 6 months);
7. Encouraging physical activity and exercise and maintaining a normal weight;
8. Safe handling, preparation, cooking and storage of food including food handling to preserve nutrients (these could have 4 sub themes and illustrations);
9. Practicing good food hygiene and sanitation such as washing hands with soap and clean water every time before preparing or eating food and at all critical times;
10. Observing portion sizes for all food groups;
11. Healthy eating for pregnant and lactating women;
12. Providing nutritious meals and snacks to school-aged children.

The research team also discussed the characteristic of sustainable diets in more detail and shared with the TWG some of the latest evidence on the "planetary healthy plate" [30] for possible inclusion in the field-testing process. Both suggestions (life cycle approach and planetary healthy plate) were agreed to. The messages were workshopped in much detail, and based on feedback from the TWG, changes were made to the text of the proposed draft FBDG messages. Messages were the same for both Mainland and Zanzibar, except for one distinct difference suggested for the vegetable message for Zanzibar and Mainland due to differences in cultural eating habits.

The final English messages were checked for translation by the TWG. These messages were deemed a final draft FBDG messages for field-testing (Table 1).

Table 1. DRAFT FBDG messages for field-testing with the population of Tanzania.

Message Nr	E = English S = Kiswahili	Message Wording
1	E	Everybody, young and old, should enjoy eating a variety of foods from different food groups every day to stay healthy and strong.
1	S	*Ujumbe: Kila mtu, mdogo na mkubwa, anapaswa kufurahia kula vyakula vya aina mbalimbali kutoka katika makundi tofauti ya vyakula kila siku ili kuwa na afya njema na nguvu.*
2	E	Add pulses such as beans, lentils, peas or nuts to your meals every day for good health.
2	S	*Ongeza vyakula vya jamii ya kunde kama maharage, choroko, njegere au karanga kwenye milo yako kila siku kwa afya njema.*
3	E	Eat animal source foods, including seafood, meat, milk or eggs every day to stay strong.
3	S	*Kula vyakula vya asili ya wanyama na baharini ikiwemo samaki, nyama, maziwa au mayai kila siku ili kuwa na mwili imara wenye nguvu.*
4	E	MAINLAND: Eat different vegetables, at least three handfuls every day to prevent and reduce risk of diseases. ZANZIBAR: Enjoy eating tomatoes, eggplant, onions, sweet pepper, carrot, okra and bitter tomatoes at least three handfuls every day to prevent and reduce risk of diseases.

Table 1. *Cont.*

Message Nr	E = English S = Kiswahili	Message Wording
4	S	MAINLAND: *Kula mboga za majani angalau viganja vitatu kila siku kuzuia na kupunguza hatari ya kupata magonjwa.* ZANZIBAR: Pendelea kula tungule, bilinganyi, vitunguu maji, pilipili boga, karoti, bamia, na nyanya chungu angalau viganga vitatu kila siku kuzuia na kupunguza hatari ya kupata magonjwa.
5	E	Eat at least two types of fruits every day for better health.
5	S	*Kula angalau aina mbili za matunda kila siku kwa afya njema.*
6	E	Eat staples such as cereals, starchy roots, tubers or plantains every day for a strong and active body.
6	S	*Kula chakula kikuu kama vile nafaka, vyakula vya mizizi (viazi, mihogo, nk) au ndizi za kupika kila siku kuwa na mwili imara na wenye nguvu.*
7	E	Limit your intake of deep fried and highly processed foods that contain fat, sugar and salt to prevent disease such as high blood pressure, diabetes and heart diseases.
7	S	*Punguza ulaji wa vyakula vilivyokaangwa na vilivyosindikwa ambavyo vina mafuta, sukari na chumvi kwa wingi ili kuzuia maradhi kama kisukari, magonjwa ya moyo, na shinikizo kubwa la damu.*
8	E	Statement: The first thousand days of a child's life starts in pregnancy and continues until two years of age. This is a very important time to ensure a child grows well on food, love and care and becomes a productive adult.
8	S	*Tamko: Siku elfu moja za mwanzo wa maisha ya mtoto huanzia wakati mama amepata ujauzito na kuendelea mpaka mtoto anapotimiza umri wa miaka miwili. Muda huu ni wa muhimu kuhakikisha mtoto anapewa chakula vizuri, kwa upendo na kujaliwa ili aweze kukua vizuri na kuwa mtu mzima mwenye nguvu na afya.*
8	E	When pregnant or breastfeeding enjoy a variety of food including animal source foods, pulses, fruit and vegetables with your meals for your health and the health of your baby.
8	S	*Wakati wa ujauzito au kunyonyesha furahia kula chakula mchanganyiko vikiwemo vyakula vya asili ya wanyama na baharini, jamii ya kunde, matunda na mboga-mboga kwenye mlo wako kwa afya yako na mtoto wako.*
9	E	Feed your baby only breast milk for the first 6 months of life and no water, herbs or porridge, because mother's milk contains all the food and water your baby needs.
9	S	*Mnyonyeshe mtoto miezi 6 ya mwanzo bila kumpa maji au uji kwa sababu maziwa ya mama yana virutubishi vyote vinavyohitajika kwa mtoto pamoja na maji.*
10	E	From 6 months feed your baby a variety of foods, including animal source foods, pulses, fruit and vegetables and continue breastfeeding up to 2 years of age and beyond for healthy growth and development of your baby.
10	S	*Mlishe mtoto wako chakula mchanganyiko, ikiwemo chakula cha jamii ya wanyama na baharini, jamii ya kunde, matunda na mboga-mboga kuanzia umri wa miezi 6 na uendelee kumnyonyesha mpaka atimize umri wa miaka 2 au zaidi kwa afya bora na ukuaji wa mwili na akili.*
11a	E	Encourage school-aged children to eat healthy snacks such as fresh fruit, vegetables and nuts.
11a	S	*Wahamasishe watoto wa umri wa kwenda shule kula vitafunwa (asusa) vilivyo bora kwa afya kama vile matunda, mbogamboga, karanga na korosho.*
11b	E	School-aged children should enjoy breakfast before school to enable them to be productive and perform well.
11b	S	*Watoto wanaokwenda shule wapewe kifungua kinywa (mlo wa asubuhi) kabla ya kwenda shule ili kuwawezesha kuwa na ufahamu mzuri na kufaulu.*
11c		Provide school-aged children with a lunchbox to take to school.
11c		*Wafungie watoto wanaokwenda shule mlo wa kula shuleni.*

Table 1. *Cont.*

Message Nr	E = English S = Kiswahili	Message Wording
12	E	Keep your food, home environment and water safe and clean to prevent diseases.
12	S	*Weka chakula, mazingira ya nyumbani na maji katika hali ya usafi na usalama ili kuzuia magonjwa.*
13	E	Make clean, safe water your drink of choice every day for good health.
13	S	*Chagua kunywa maji safi na salama kuwa kinywaji chako kila siku kwa afya njema.*
14	E	Avoid drinking alcohol and tobacco use to prevent and reduce risk of diseases.
14	S	*Epuka matumizi ya pombe na bidhaa zitokanazo na tumbaku ili kuzuia na kupunguza uwezekano wa kupata magonjwa.*
15	E	Be physically active every day to stay strong and keep a healthy weight.
15	S	*Ushughulishe mwili wako kila siku ili kuwa imara na uzito sahihi kiafya.*

2.2. The Development of Images to Accompany FBDG Messages—Artwork Process

The research team aimed to develop images that were representative of the context. The graphic designer in the team led this process using mixed imagery, i.e., drawn images of persons and/or settings with photographs placed in the images for real and accurate representation of foods. To inform this process, during various meetings, the researchers asked pertinent questions to the FAO technical team and the TGW about the local food environment, foods that are culturally acceptable and available, and what should be included in the images. To create the artwork for the images of the FBDG messages, reference photographs were taken of food environments (e.g., food markets, retail stores), indigenous and locally consumed foods from all food groups, some tableware and a home environment. No persons were photographed.

Visits to local food markets and shops/supermarkets, accompanied by the FAO technical team and members of the TWG, were undertaken to obtain the required foods. Raw and cooked versions of foods, with branding removed if applicable, were photographed in a lightbox to minimise shadows and obtain a clear, well-lit photograph of the food, whole and in portioned amounts. These photographs were reworked and edited to create photographs which could be used in the images. Once the written guidelines were established, images were digitally drawn to illustrate them, placing the photographed food where necessary and appropriate.

2.3. Design and Setting

A qualitative study was conducted with focus group discussions as a data collection method. Qualitative research methods are most useful in understanding the viewpoints of individuals regarding sensitive and socially dependent concepts [31]. The study sub-sites (municipalities) were purposively selected in consultation with FAO and the TWG. The individuals represented in these groups collectively have extensive country-wide experience and attempted to include as much of the diversity of the population in the selection of the sites. The following sites were decided upon: (1) Dar es Salaam, (2) Mwanza and (3) Iringa Municipalities for the Mainland and (4) Kusini for Zanzibar. Dar es Salaam and Kusini represented urban sites; Mwanza and Iringa represented rural sites (Figure 1).

2.4. Study Population

Women were chosen as the study population as they are the primary family caregivers and generally make decisions regarding food choices and preparation. They may or may not receive support from partners/men, but they are the primary group who require empowerment to make the right decisions with regard to nutrition. Inclusion criteria were: women of 18 years and older, who speak and understand Kiswahili, and who were permanent residents of the chosen study sites. Exclusion criteria pertained to women who did not provide informed consent for participation, who had formal training in nutrition and those unable to participate in the study due to limited mental abilities or comprehension.

2.5. Sample Size

Sample size was determined by data saturation—a recognised qualitative research approach used where an investigator extracts data until sufficient information has been collected or until no new information can be obtained [32]. Three to four focus group discussions (FGDs) were planned in each of the four study sites (12–16 FGDs), consisting of 6–8 participants each, as per the recommendations by the FAO [1]. However, it was felt that 14 messages were too many to test in one FGD. The messages were therefore split so that 7 messages were tested per FGD with an anticipated 24–32 FGDs that could be conducted.

Sampling Strategy

Non-random purposive sampling was used to select participants for the FGDs, which is the appropriate sampling technique used for qualitative research [32]. Purposive sampling includes the process of intentionally selecting participants for the study based on certain characteristics [32].

In each selected study site, fieldworkers recruited potential individuals from households and various community-based settings. All of the identified participants were provided with information regarding the study. They were given a date and time of the FGDs and contacted by telephone or a visit from a fieldworker prior to the FGD, to promote attendance.

2.6. Training of Fieldworkers

As agreed between FAO and the TWG members, the TWG members (25 in total) acted as fieldworkers for this research study. The research team conducted FGD facilitation skills training, as well as feedback and reflection as part of pilot testing over a three-day period with the fieldworkers in August 2019 to embed theory and gain practical experience. Each fieldworker was supplied with a training manual to explain all the research processes. Training was presented in English, since all TWG members can speak, write, and understand English. The training sessions included an overview of the study's aim and objectives as well as the inclusion and exclusion criteria and methodology. Fieldworkers were required to be familiar with all research documentation as well as the recruitment form which was used in the field to recruit participants.

The fieldworkers were thoroughly trained on the FGD guide. The FGD guide was translated into Kiswahili by trained professionals then back translated into English for quality control purposes. The role of the FGD guide was to provide the fieldworkers with direction while facilitating all discussions. Fieldworkers were also trained to take notes during FGDs regarding the dynamics of the group as well as body language and facial expressions of the participants. All these aspects were covered in the training session.

Since the final ethics clearance for the project was only received from the Zanzibar Health Research Institute by the time of pilot testing, and not from the Mainland Dar es Salaam office, all training and pilot pre-testing took place in Zanzibar on 23 August 2019.

2.7. Pilot Testing

Pilot field-testing was conducted as part of the fieldworker training and served two purposes. Firstly, it provided an opportunity for testing and validating content (messages and images), therefore gauging usefulness and relevance for the Tanzanian context. Secondly, it provided the fieldworkers with hands-on practical training on how to conduct FGDs using the FGD guide. Eleven "mock" FGDs were conducted. Fieldworkers worked in teams and transcribed the audio recordings. Four of the authors analysed the transcripts. This process assisted the research team to identify areas where further training was required, thereby consolidating learning and ensuring fieldworkers were prepared and confident to conduct the upcoming FGDs for the main study in their various regions.

From the pilot test findings, it was clear that the infant and young child feeding (IYCF) messages were already in use and well understood and communicated widely. In agreement with the TWG and FAO, a decision was therefore made to adopt the IYCF messages as detailed in Table 2 (messages 8–10) without needing further field testing.

Table 2. Sociodemographic characteristics of participants.

Variable		n (%) $^\alpha$
District/region	Dar es Salaam	96 (29.6)
	Mwanza	98 (30.5)
	Iringa	72 (22.2)
	Kusini (Zanzibar)	58 (17.9)
Age	* Average age in years (Mean ± SD)	35.01 ± 11.62
Women with children	Yes	277 (85.5)
	No	47 (14.5)
Number of children	* Average number of children (mean ± SD)	3.22 ± 2.30
Ages of children	* Average age of children in years (mean ± SD)	13.93 ± 10.48
Mother/caregiver of children under 5 years of age	Yes	174 (53.7)
	No	150 (46.3)
Home Language	Kiswahili	322 (99.4)
	* missing data	2 (0.6)
Education status	None	15 (4.6)
	Primary school/Grade 1–7	196 (60.5)
	Secondary school/Grade 8–10 (Form i–iv)	107 (33.0)
	Secondary school/Grade 11–12 (Form v–vi)	2 (0.6)
	Diploma	2 (0.6)
	Tertiary education (college, university)	0 (0.0)
	* missing data	2 (0.85)
Employment status	Employed	193 (59.6)
	Unemployed	76 (23.5)
	Unemployed, not looking for work	55 (17.0)
Role relating to food in the household	Provide or contribute money for food	248 (76.5)
	Decide what food should be bought or used in the house	240 (74.1)
	Purchase food	252 (77.8)
	Prepare food	302 (93.2)
	Grow food for use in household	212 (65.4)
	Grow food and sell produce for money	158 (48.8)
	Other	17 (5.2)

$^\alpha$ Percentage calculated from total sample N = 324. * Averages given in mean and standard deviations (mean ± SD).

In preparation for fieldwork, consent forms and questionnaires were translated into Kiswahili by trained professionals and translated back into English for quality control purposes.

2.8. Focus Group Discussions

Fieldwork took place in Zanzibar from 29 October 2019 to 3 November 2019 and on the Mainland from 25 November to 3 December 2019. Before commencement of the FGDs, all participants were requested to complete an informed consent form followed by a short socio-demographic questionnaire. The questionnaire consisted of information such as the participant's date of birth, home language, highest level of education, employment status and role in the household pertaining to food. Fieldworkers guided the participants through the questionnaire by reading out all the questions and ensuring participants completed questionnaires correctly. The questionnaire was completed by a fieldworker in the case of a participant being illiterate.

All FGDs were approximately two hours in duration and were conducted at local libraries, town halls, school classrooms or suitable facilities at other community-based organisations. Facilities were contacted or visited in advance to arrange suitable dates and times for the FGDs. Venues were set up on the day of the discussion and refreshments provided. All FGDs were audio recorded, with consent from participants, for analysis purposes.

Each FGD was conducted by trained fieldworkers working in pairs. The FGD was guided by a discussion guide, which was adapted from similar studies (Supplementary Materials) [27,33]. The discussion guide included the procedure that had to be followed, and questions to be asked and prompts that could be used to stimulate discussion on the proposed FBDGs and images. These included discussion points on previous exposure to similar messages, the mother's/caregiver's understanding and interpretation of messages and images, the perceived importance of the messages and images, as well as barriers and enablers to following of the messages. Participants were also asked how they would reword the message to make it more understandable to the general public. The guide was translated from English into Swahili by trained professionals. The document was back-translated into English for quality-control purposes. The same procedure was used for the translation of the socio-demographic questionnaires and consent forms.

The proposed FBDG messages and images were printed on flash cards and A3 size posters to assist discussions. FGDs began with fieldworkers welcoming the participants and explaining the main aim and expectations of the study. The fieldworkers read out one proposed message at a time, presenting a flash card of the proposed FBDG message, and facilitating discussions among the participants. The same format was used for the images. The proposed FBDG messages and images were numbered for field testing purposes, so that audio recordings could accurately reflect the message/image under discussion.

2.9. Data Analysis

After completion of FGDs, fieldworkers sent all audio files to the research team as detailed in the research protocol via "We Transfer", an Internet-based computer file transfer service. The research team in turn forwarded the audio files to professional transcribers in Dar es Salaam, also via "We Transfer", for verbatim transcription in Kiswahili followed by translation into English.

Three of the authors performed manual content analysis on the transcribed English translations. Each interview was assessed by at least two of the three authors. They independently inspected the data for common themes and coded the data, predominantly by deductive coding, after careful reading and re-reading of the text. Themes were organized according to broader clusters, based on the predetermined study objectives and the discussion guide. They discussed the analyses and reached consensus on the interpretation of the findings, as reported in the results section.

3. Results

3.1. Socio-Demographic Profile of Participants

Ten FGDs were conducted for Zanzibar (n = 58 participants) and 24 FGDs were conducted for Mainland (n = 266 participants), resulting in a total of 324 female participants. The FGDs were conducted with on average 6–8 (up to a maximum of 11) participants per group. The average age of participants was 35.01 ± 11.62 years, predominantly Kiswahili speaking (n = 322; 98%), and employed (n = 193; 59,6%) with 60,5% (n = 196) reaching at least a primary level of education. Most participants (n = 277; 85,5%) had children with an average age of 13.93 ± 10.48 years. Of the participants, 174 (53.7%) had children under the age of five years. Participants fulfilled various roles pertaining to food in the household, including deciding what food should be bought/used (*n* = 240; 74%), purchased (*n* = 252; 78%) and prepared (*n* = 302; 93%), growing food for own household use (*n* = 212; 65%) and selling produce for income (*n* = 158; 48,8%) (Table 2).

3.2. General Understanding of the Proposed FBDG Messages and Images, and Recommendations for Improvement

In general, the FGD participants in Zanzibar and Mainland understood the core meaning that the FBDG messages and images intended to convey. Participants reported that they had been exposed to similar messages mostly through radio, television, clinics/hospitals and schools. These were also the most suggested platforms for distribution of the final FBDG messages and images.

> *"We need to use such foods so that we can put our body into a good health and our children to have strong minds and grow physically and mentally fit."* (Participant 2 ZG2)

Participants reported that the messages and images complemented one another, and that the community would mostly be able to understand them as well. However, they did feel that education would be required for improved comprehension of the messages.

> *"Yes, they will understand when they are exposed to education."* (Participant 6 ZG2)

The messages aimed at school-aged children were not as familiar to participants and they expressed a need for more education on these messages specifically. During discussion, participants expressed appreciation for these messages, particularly making the link between a well-fed child being more receptive to education.

> *"What I know is that in this message we are supposed to mobilize our children to eat food that will help build their health and also maintain their mental capability so as to manage well in their studies."* (Participant 1 ZG3)

The messages and images on water, sanitation and hygiene (WASH) were very well known and it was clear that these had been widely communicated and distributed through government education initiatives.

Some suggestions were made to make messages and images more culturally appropriate, comprehensive, and understandable. Suggestions for cultural appropriateness were made in the Zanzibar context to cover women's breasts in the breastfeeding image, cover girl's legs in the school meal image, and separate men and women in the activity image. To make the messages and images more comprehensive and understandable, participants suggested adding specific food products or more examples of a specific food group (e.g., different fruits, vegetables and meats). A specific comment was made to label individual foods in certain images with text. However, when images were discussed, participants could clearly identify the foods, and since images are intended to stand independently, this suggestion was not incorporated in the final set of proposed FBGD messages and images.

3.3. Specific Understanding of the Proposed FBDG Messages and Images, and Recommendations for Improvement

In this section, each FBDG message is stated, and image portrayed, followed by a description of findings, specific suggestions for change and/or additions and deletions to the messages and images. Following on, the revised image and message is depicted, as relevant, Figures 2–19 (copyright permission has been obtained from FAO and the artist (@Fao/Shân Fisch-er)).

Original Image

Figure 2. Everybody, young and old, should enjoy eating a variety of foods from different food groups every day to stay healthy and strong.

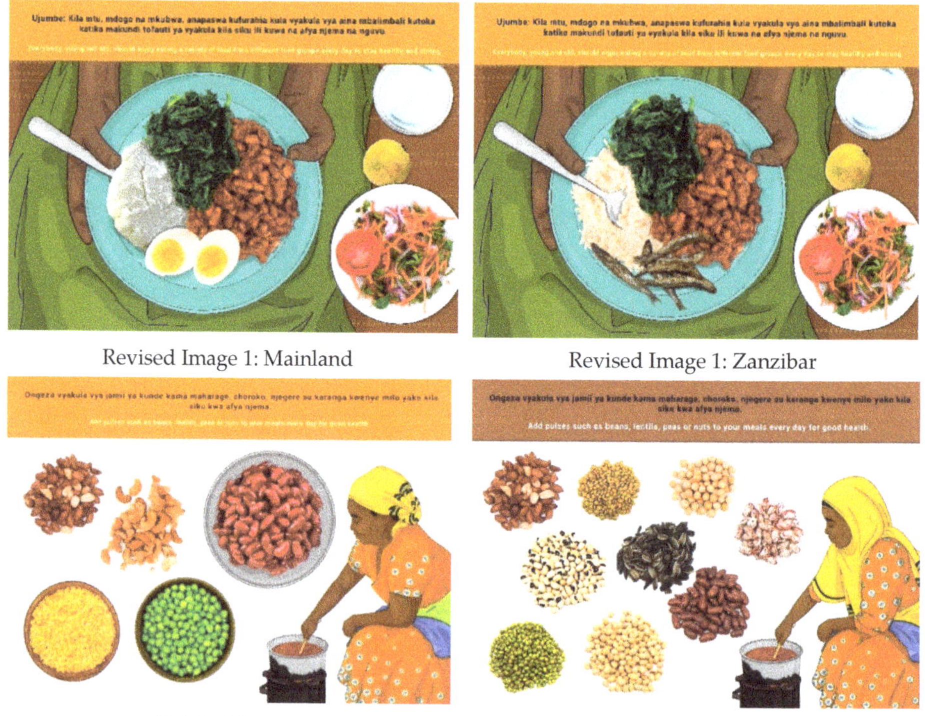

Revised Image 1: Mainland

Revised Image 1: Zanzibar

Original Image

Revised Image: Mainland and Zanzibar

Figure 3. Add pulses such as beans, lentils, peas or nuts to your meals every day for good health.

Original Image   Revised Image: Mainland and Zanzibar

Figure 4. Eat animal source foods, including seafood, meat, milk or eggs every day to stay strong.

Original Image Mainland   Original Image Zanzibar

Figure 5. Eat different vegetables, at least three handfuls every day to prevent and reduce risk of diseases.

Revised Image: Mainland and Zanzibar

Figure 6. Revised message and image 4 for Mainland and Zanzibar: Eat different coloured vegetables such as tomatoes, onions, carrots, okra and green leafy vegetables every day to prevent and reduce risk of diseases.

Original Image   Revised Image

Figure 7. Eat at least two types of fruits every day for better health.

Original Image   Revised Image

Figure 8. Eat staples such as cereals, starchy roots, tubers or plantains every day for a strong and active body.

Original Image   Revised Image

Figure 9. Limit your intake of deep fried and highly processed foods that contain fat, sugar and salt to prevent disease such as high blood pressure, diabetes and heart diseases.

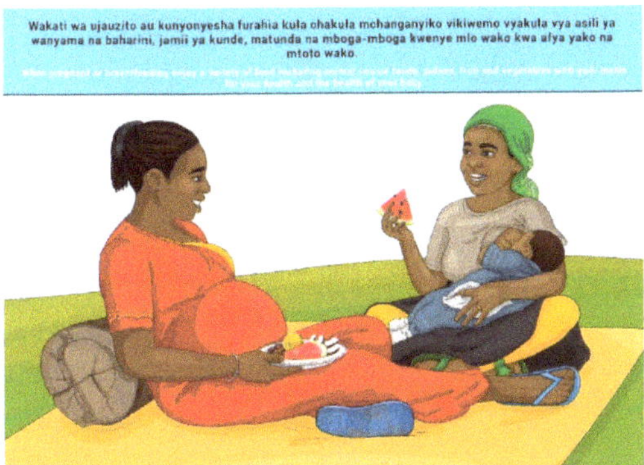

Figure 10. When pregnant or breastfeeding enjoy a variety of food including animal source foods, pulses, fruit and vegetables with your meals for your health and the health of your baby.

Figure 11. Feed your baby only breast milk for the first 6 months of life and no water, herbs or porridge, because mother's milk contains all the food and water your baby needs.

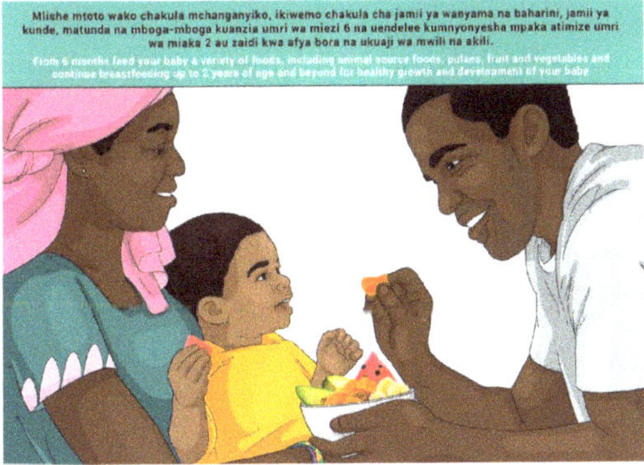

Figure 12. From 6 months feed your baby a variety of foods, including animal source foods, pulses, fruit and vegetables and continue breastfeeding up to 2 years of age and beyond for healthy growth and development of your baby.

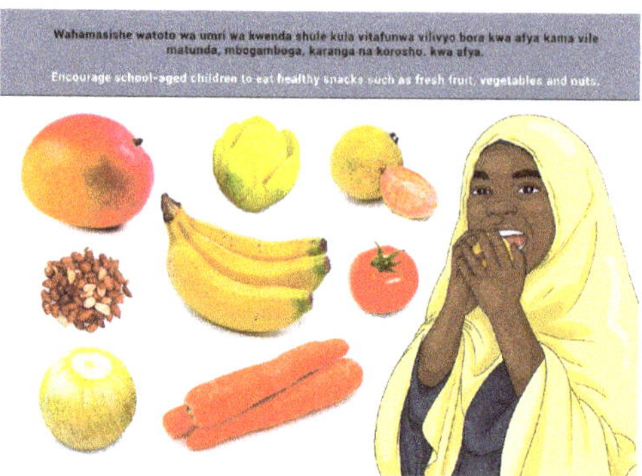

Figure 13. Encourage school-aged children to eat healthy snacks such as fresh fruit, vegetables and nuts.

Figure 14. School-aged children should enjoy breakfast before school to enable them to be productive and perform well.

Figure 15. Provide school-aged children with a lunchbox to take to school.

Original Image

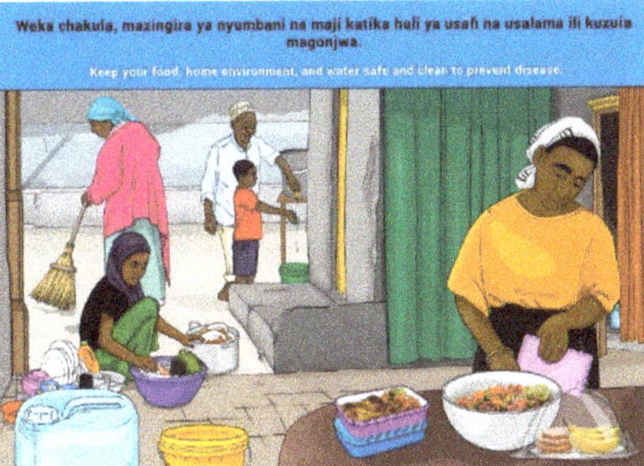

Revised Image

Figure 16. Keep your food, home environment and water safe and clean to prevent diseases.

Original Image

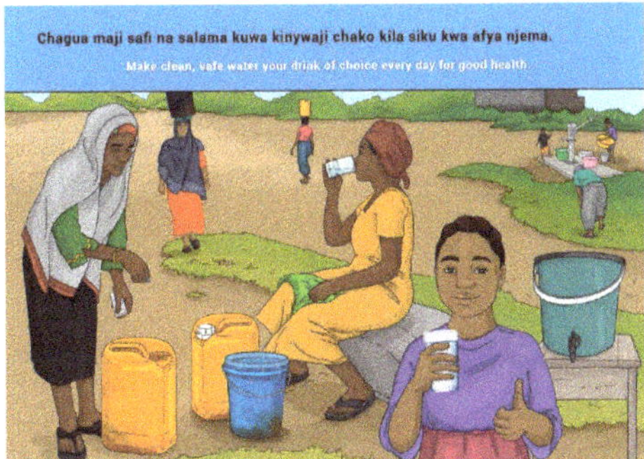

Revised Image

Figure 17. Make clean, safe water your drink of choice every day for good health.

Figure 18. Avoid drinking alcohol and tobacco use to prevent and reduce risk of diseases.

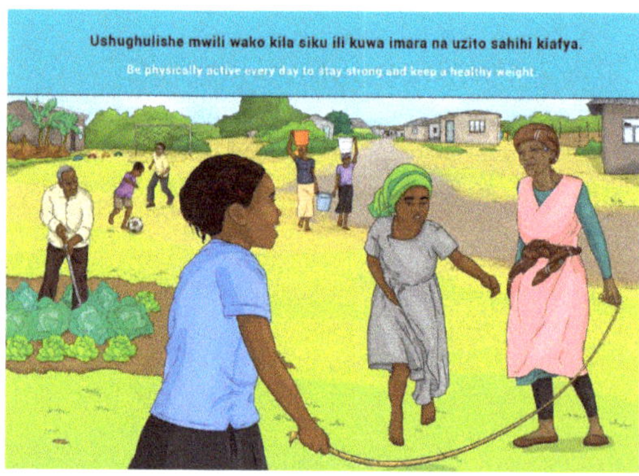

Original Image   Revised Image

Figure 19. Be physically active every day to stay strong and keep a healthy weight.

The message and image elicited good understanding from participants. The image supported understanding of the word "variety", including that the food is "nutritious", "healthy", "balanced "/" mixed foods" and "good food". In some instances, participants responded that not all members of the public will be able to understand and implement the message and image, due to poverty and the resultant inability to purchase some products.

"Due to high levels of poverty in the villages, some members of the public will not understand this picture . . . " (Participant 6 MM2)

Suggestions were made that the fluid in the image should be changed from juice to tea. The researchers suggested that water should be depicted in the glass to correspond with and enhance message and image 13. Since green leavy vegetables are commonly consumed, the green beans were removed and replaced with spinach. A recommendation was made from Mainland participants to replace the rice on the plate with *ugali* (stiff, cooked porridge) since *ugali* is considered the most consumed staple food in the country. Zanzibar participants, however, could relate to the rice on the plate. In addition, Zanzibar also favoured small fish (sardines) to the boiled egg.

The message and image were well understood, and no changes were suggested. Some discussion about beans causing ulcers occurred. This misinformation was apparently received as advice from doctors at hospitals and may be linked to the suggestion to avoid these foods in the case of irritable bowel syndrome (IBS). Since FBDG are aimed at the healthy population, this message was not altered to incorporate this specialised dietary advice.

A review of the term "nuts" in the Kiswahili message was suggested since there seemed to be ambiguity around the term. Some participants interpreted it as cashew nuts, and the image also depicted it. Most participants deemed cashew nuts as difficult to access or expensive. The image of cashew nuts was therefore removed, and since there is a wide variety of different pulses available in the country, more varieties were added to the revised image.

The message and image were well understood, and reference was made to "body building foods". Animal source foodsi n Zanzibar mainly comprise seafood, whereas on the Mainland it is freshwater fish from the lakes and other animals. Participants from the Mainland could therefore not relate to the regular consumption of seafood, specifically octopus. A recommendation was made by Mainland FGDs to add red meat to the image. The image was changed to portray a wider variety of products (freshwater fish, octopus, red meat on a skewer as well as cooked chicken).

Indigenous foods, such as local insects, were recommended in a few instances in the Mainland FGDs. However, it is important to note that insects are generally region specific and would therefore require such consideration when included in education.

This message and the image were well received but elicited a lot of debate. The wording of "handfuls" was not well understood. Other methods of showing portion size were therefore suggested, such as "spoonfuls" made by Mainland FGDs or "bowls". It was also suggested that frequency ("three times a day" and "included in three meals a day") or listing the different types of vegetables by name.

There was a recommendation from Zanzibar to include okra in the image. Mainland suggested that green leafy vegetables, such as sweet potato leaves, cassava leaves and spinach should be included in the image as well.

The message was well understood, and no changes were suggested. Both Zanzibar and Mainland participants suggested adding more fruit such as pineapples, pawpaw, mangoes and avocados to the image. In Mainland, oranges were deemed difficult to access, so the suggestion was made to remove this fruit from the image.

In some instances, the message and image were incorrectly interpreted as "a variety of food". *Ugali* is the most consumed staple on Mainland, and it was suggested to be depicted clearly in the image. Most participants could not identify the sweet potatoes in the image, and therefore, a variety of tubers were suggested for inclusion in the image. "Chapati" (a thin pancake of unleavened whole-meal bread cooked on a griddle) was removed from the original image to create space for the tubers.

The message and image were well understood; however, the need for education on the specific diseases mentioned were expressed. There was a general understanding that these foods are found in supermarkets and should be avoided. There was a recommendation to add biscuits, more hard-boiled sweets, and sugar sweetened drinks to the image.

Maternal, Infant and Young Child Feeding (MIYCF) messages and images.

The MIYCF messages were introduced with the following pre-amble statement: The first thousand days of a child's life starts in pregnancy and continues until two years of age. This is a very important time to ensure a child grows well on food, love and care and becomes a productive adult.

There are three messages and images suggested for the MIYCF period.

The MIYCF messages were already in use and well understood and communicated widely, and no further testing was conducted for understanding of the messages. Suggested changes to the images were related to religious aspects of covering of pregnant and breastfeeding women's faces and bodies for Zanzibar specifically. These changes were incorporated in the final set of images for Zanzibar (please see Section 3.3).

The message and image were well understood. However, in discussion of this message, some issues were raised in reference to available funds to afford healthy snacks. The suggestions made for changes to an accompanying image were as follows:

○ Image with children around a table depicting peanuts and vegetables;
○ A child in school uniform being handed the suggested snacks by an elder or father;
○ A child standing next to the suggested snacks.

Since the image was already in the format of the latter suggestion, no further changes were made to this image.

A few participants raised concerns about the availability of money to buy healthy food in general, but also in reference to the message on supplying breakfast to children before school. It was mentioned that preparing breakfast is taxing on caregivers and that parents would need motivation to follow this message. This is due to limited time availability in the mornings.

Participants required clarity on what the woman was cooking in the image; therefore, the consistency of the porridge was changed to portray a "grittier" texture in the revised image. Another suggestion for change to the image included a mother providing tea and *chapati* to her children. It was also recommended that fruit be added to the image.

In Zanzibar, participants requested that the head and body of the women cooking should be covered. These changes were incorporated in the final set of images for Zanzibar (please see Section 3.3).

This message was generally not well understood by Mainland FGDs. Participants associated 'school-age' with younger children up to primary school rather than children of all ages attending school. It was felt that implementation would be difficult because most parents leave home early to go to work, but mostly because packing a lunchbox for children was unfamiliar as they receive porridge at school through the school nutrition programme. This was considered a better approach to reducing inequality and avoiding stigmatizing children whose parents may not be able to afford to pack a lunchbox.

> *"The message should insist on lunch at school to avoid inequality at school which can create problems for the poor children."* (Participant 1 MD2)

Children are not encouraged to bring lunch to school by teachers, and parents were uncertain that children would be able to take care of packed lunches. Participants were especially worried that giving their children packed food will lead to fights/bullying at school. Participants also discouraged giving money to buy food at school/tuck-shop.

> *"I would draw a picture showing all the children having lunch at school and not packed lunch which may create conflicts."* (Participant 3MI1)

Suggestions for changes to the image included:
- A parent provides a child with a safe storage container containing food for school;
- A child washing hands with a lunchbox depicted in close proximity;
- Add a school building to the background;
- Add a tap for washing hands;
- Add a shared plate so that it represents a school feeding scheme.

A combination of porridge, depicting the school nutrition programme and shared lunchboxes were included in the revised image.

The message was well understood, but the image stimulated a lot of debate. The discussions were related to the man in the picture washing dishes. It was interpreted that if a man is washing the dishes/utensils, then he is not working and earning an income for the household. It was suggested that the man should be replaced by a woman, although some participants felt that the man could be depicted. Another suggestion was the inclusion of a pile of clean dishes in the image.

> *"It will not be understood because men do not work in the kitchen. The woman should be in the picture instead and more food should be included."* (Participant 6 MI1)

Participants also felt that the image looked too modern and proposed that it should show a typical Tanzanian environment including the inside and outside of the house where both men and women should be engaging in cleaning activities. Other suggested changes included depicting clean water, safely stored food, and sweeping of the floor.

> *"The picture should show a clean house, with a clean kitchen, and the hair of the cook must be covered, and the other picture should show a house in a dirty environment and food covered with flies, for comparison."* (Participant 3 MI1)

As all the other FBDG images feature women (e.g., preparing and cooking food), the research team felt that this was the most appropriate image to include a man sharing the workload of domestic tasks. The image was therefore reworked to show a man and a young boy collecting water from an outside tap.

This message was well understood, but the focus was rather on the hygiene aspect of clean water rather than making it the drink of choice daily. The message therefore needs to be accompanied with education on increasing intake of clean, safe water. A suggestion was made to indicate water boiling on a stove or water being purified, and/or a bucket showing that the water is clean/has been purified. In the revised image, activities around water collection and purification were portrayed. The man drinking bottled water was also replaced by a woman drinking water from a glass.

Participants articulated the meaning of both the message and the image very well. They commented that the image complemented the message well. A concern was raised that the government allows the trade in alcohol and cigarettes, hinting that the government and the proposed FBDG message are saying the opposite, which are two conflicting policy positions of government (health and trade policy).

Participants proposed there should be a person affected by the consumption of alcohol and cigarettes depicted in the image. The researchers discussed this suggestion and felt that the image should support the word "avoid" and should reflect a positive action of saying "no" to these products. No further changes were therefore made to the image.

This message was well understood, but quite a few suggestions were made to improve the acceptability of the image, including to show different kinds of activities, e.g., agriculture/gardening, a woman carrying water, skipping rope, a game such as "nage" (ball game) and swimming. Another suggestion was made to separate men and women while doing any activity and to cover women's legs and arms, particularly by Zanzibar participants. In some of the Mainland FGDs, there was discomfort with women and the elderly engaging in physical activity. Therefore, this message should be accompanied by education about the benefits of physical activity for all age groups.

3.4. Final Considerations

Suggestions for changes to the proposed FBDG messages and images were incorporated as far as feasible.

During a pre-final online workshop in early 2021 between the researchers, TWG and FAO, the representatives from Zanzibar requested that all the applicable images be tweaked to portray Muslim dress specifications. It was eventually decided to prepare a separate set of images for Mainland and Zanzibar with the distinction of Muslim dress for all the images intended for Zanzibar.

The research team suggested that Message 1 (Everybody, young and old, should enjoy eating a variety of foods from different food groups every day to stay healthy and strong) can be used as an over-arching message for the FBDG messages, and Image 1 could be retained to depict a planetary healthy plate for Tanzania.

These suggestions were put forward to the TWG from Mainland and Zanzibar for their final consideration.

4. Discussion

This research study aimed to determine the understanding, acceptability, and feasibility of proposed FBDG messages and images amongst consumers in Tanzania. The importance of nutrition throughout the life cycle [28,29] combined with the value of focussing on food as the single strongest lever to optimize human health and environmental sustainability [30], inspired the researchers and the TWG to adopt a lifecycle approach with a focus on sustainable eating in the development of the proposed messages and images for the food-based dietary guidelines of the country.

It was evident from the findings of this study that in general, participants understood the meaning of the proposed FBDG messages and found them to be feasible, bearing in mind that a large proportion of the population live in poverty [9]. Furthermore, the images accompanying the messages were mostly well received and regarded as acceptable. Participants made practical suggestions for improved applicability and understanding, which was incorporated in as far as feasible to the final proposed set of FBDG messages and images for Tanzania. Since the images will be copied, sometimes in black and white, the graphic designer was mindful of the amount of detail that could be changed or added to certain images. Too much detail could distort the images when copies are made, and the focus of the image could be lost.

Although participants displayed some knowledge of dietary concepts, misinformation and lack of information on some aspects of nutrition were evident. In particular, the message on staple foods (Message 6: Eat staples such as cereals, starchy roots, tubers or plantains every day for a strong and active body.) was interpreted as "a variety of food" by some participants. This could be due to a diet based on staple foods and prevailing poor dietary diversity in the country [22]. These finding are consistent with widespread financial constraints, poor household food security and low socio-economic status in many Sub-Saharan countries, despite some degree of economic growth, and political and social transition [34]. The COVID-19 pandemic is expected to exacerbate undernutrition and food insecurity due to, among other factors, the vulnerability and weaknesses of already fragile food—and strained healthcare systems. Inequities in food and health systems worsen inequalities in nutrition outcomes that, in turn, can lead to more inequity, fuelling a vicious cycle [34]. It is therefore more important now than ever to scale-up efforts to support and educate the public about healthy and sustainable diets. In the context of the current study, the overarching message (Everybody, young and old, should enjoy eating a variety of foods from different food groups every day to stay healthy and strong) should be widely communicated and promoted, to ensure understanding of the staple message and image in particular.

An encouraging finding is that participants realised the need and expressed a desire for nutrition education to accompany the dissemination of the FBDG messages and images. It has been shown that social and behaviour change communication can effectively improve nutrition knowledge [35]. This could include interpersonal (counselling, education support groups), mass media (television, community radio, printed media, social and/or mobile technology) and community mobilization efforts (health days, campaigns) and corresponds with the suggestions made by participants for dissemination efforts.

Most studies on healthy diets and sustainable food production agree that a diet rich in plant-based foods with fewer animal source foods is good for human health and the environment [30]. A planetary healthy plate should consist of half a plate of vegetables and fruits and the other half should contain whole grains, plant protein sources, unsaturated plant oils, and (optionally) small amounts of animal protein sources. However, it is recognised that some populations depend on growing crops and raising livestock [30]. In addition, many populations suffer from high burdens of undernutrition and struggle to meet micronutrient needs from plant source foods alone. Evidence suggests that infants and young children also need animal source foods to grow and develop. Given these considerations, the role of animal source foods in people's diets must be carefully considered in each context alongside local and regional realities [30]. It was therefore recommended that the "planetary healthy plate" image, field tested with Message 1, could be retained to convey information on sustainable eating to consumers in Tanzania.

According to a global review of FBDGs from 90 countries, the food groups most adhered to were the starchy staples (e.g., rice and potatoes) and fruit and vegetables [36]. Adequate intakes of fruit and vegetables are associated with a reduced risk of chronic diseases and body weight management. The WHO and FAO recommend that adults consume at least five servings of fruits and vegetables per day excluding starchy vegetables. Regardless of the evidence showing protective effects of fruits and vegetables, intakes are still inadequate in many countries, especially LMICs. For this reason, public health efforts and improved strategies, notably well-planned and behaviour-focused nutrition education intervention to promote fruit and vegetable intake, are critical [37]. Differences in names and cooking methods for certain foods on the Mainland and in Zanzibar, particularly for vegetables, made wording of the message challenging. For example, in Zanzibar, vegetables such as carrots, onions and tomatoes are referred to as spices. After much deliberation in the TWG, the vegetable message and image were finalized as follows: "Eat different coloured vegetables like tomatoes, onions, carrots, okra and green leafy vegetables every day to prevent and reduce risk of diseases". The images were adapted to depict the most consumed, culturally acceptable and available products.

From the eight countries in Africa with published FBDGs (Benin, Kenya, Namibia, Nigeria, Seychelles, Sierra Leone, South Africa and Zambia), South Africa [27] and Zambia have a set of Paediatric FBDGs, and the Seychelles revised 2020 version has a section on IYCF. It was therefore deemed imperative that the FBDGs and images for Tanzania should include messages and visual illustrations to portray the importance of optimal nutrition during the first 1000 days of life [28,29]. According to the Tanzanian National Nutrition Survey (2018), 96.6% of children 0–23 months were ever breastfed and almost 58% of infants younger than six months were exclusively breastfed. This is an improvement from 2014 (41.1%). In Zanzibar, a significant increase in exclusive breastfeeding was observed between 2014 and 2018 (19.7% to 30.0%). Lower levels of exclusive breastfeeding on the Island are due, among other reasons, to religious practices at birth (e.g., smearing honey in the newborn's mouth and giving water). At national level, it was reported that 86.8% of children aged 6 to 8 months were timeously introduced to complementary food, but dietary diversity remains sub-optimal [13]. A study by Mgongo et al. (2014) [38] revealed that although mothers understood the concept of exclusive breastfeeding and were positive about the practice, there were many barriers that mothers faced. Among the barriers were poor support and conflicting advice from influential people, including mothers-in-law, friends and healthcare workers. Returning to work and the pressures of earning an income was also highlighted as a barrier. Similar findings have been reported elsewhere on the continent, notably East Africa (Tanzania, Ethiopia, Mozambique and Kenya) [39] and South Africa [40]. The Tanzanian government has invested in nutrition education at primary health care level, and it was clear from the pilot testing of the FBDG messages that IYCF messages have been widely communicated. These activities as well as more support for working mothers, should be strengthened and continued to further improve IYCF practices in the country.

Eating breakfast is recommended due to its association with improved macro- and micronutrient intakes, BMI and lifestyle. Breakfast is also widely promoted, particularly for school-aged children and adolescents, to improve cognitive function and academic performance [41]. School nutrition programmes aim to address short-term hunger and nutrient deficiencies, improve school attendance and performance, and support local agriculture and the economy [42]. School nutrition programmes further provide benefits for the physical, mental, and psychosocial development of school-age children and adolescents, particularly in LMICs [43]. In the current study, the proposed FBDG messages and images aimed at school-aged children elicited a lot of debate. Barriers that were mentioned to implementation of these messages included time constraints for parents to prepare breakfast and lunch packets/boxes, poverty, fear of bullying of children who take food to school and the fact that public schools in Tanzania have a nutrition programme in place. In the light of challenges worsened by the COVID-19 pandemic, governments have in some cases been forced to reduce social support, such as school nutrition programmes [34]. The most vulnerable and marginalised individuals and groups often rely on these support initiatives and there is a real risk that, as nations try to recover from the pandemic, the gains that were made pre-COVID in reducing hunger and malnutrition may be lost [34]. For these reasons, the promotion of eating breakfast and taking a lunchbox to school, if possible, remain important, as well as provision of school meals, particularly in LMICs.

Limitations

Differences in names and cooking methods for certain foods on the Mainland and in Zanzibar made wording of the message and images challenging. Cultural differences, especially in clothing and dressing also made it difficult to have one representative image for a concept for the Republic of Tanzania (Mainland and Zanzibar). This was resolved by the final decision between the research team, TWG and FAO to prepare separate sets of FBDG messages and images with a few distinct differences in clothing/dressing and certain foods.

5. Conclusions and Recommendations

The proposed FBDGs messages and visual illustrations for Tanzania have been field-tested for understanding, acceptability and feasibility. The study results indicate a general awareness of the messages, but some rewording of certain messages and editing of certain images were suggested to facilitate the comprehension of the message and image. The

TWG for Zanzibar FBDGs and Mainland FBDGs incorporated the messages and images in the respective FBDGs Technical Manuals that were compiled in the last quarter of 2021 and are due to be launched in 2022. It is recommended that the field-tested FBDG messages and images with the incorporated suggested changes be adopted to form part of the national nutrition education efforts by the Tanzanian government.

Once adopted, the FBDGs messages and images should be used to educate various stakeholders, including parents, caregivers, healthcare providers and educators on appropriate nutritional practices for children and adults. The messages can be used to guide implementation of nutrition education to promote healthy diets while simultaneously addressing the abundance of misinformation on food and nutrition. The use of FBDGs can ensure consistent messages to support the healthy growth and development of young children in Tanzania as well as general nutritional well-being of adult consumers in the country.

It is also strongly recommended that FBDGs should be applied more broadly and not just within food and nutrition education programmes. Experts propose a food-systems approach whereby FBDGs should be incorporated into programmes, policies and other publications of national and sub-national departments in sectors such as agriculture, social development and safety and security in order to contribute to the achievement of sustainable healthy diets. Once implemented, the use and impact of the FBDGs messages should be monitored and evaluated [44].

Supplementary Materials: The following are available online at https://www.mdpi.com/article/10.3390/nu14132705/s1, Addendum 1: FOCUS GROUP DISCUSSION SESSION OUTLINE.

Author Contributions: Conceptualization, L.M.D.P., M.P.C. and M.A.; methodology, L.M.D.P. and M.P.C.; formal analysis, L.M.D.P., N.J., A.C. and S.F.; data curation L.M.D.P., N.J., A.C., S.F. and M.P.C.; writing—original draft preparation L.M.D.P., N.J., A.C., S.F. and M.P.C.; writing—review and editing, L.M.D.P., N.J., M.P.C., A.C., S.F., M.A. and P.L.; project administration, L.M.D.P. and M.P.C.; funding acquisition, L.M.D.P., M.A. and M.P.C. All authors have read and agreed to the published version of the manuscript.

Funding: This research was funded by the Food and Agriculture Organisation under a Letter of Agreement (LOA). Stellenbosch University, the Division of Human Nutrition, were appointed by the United Nations Food and Agriculture Organisation (FAO) Sub-Regional Office for Southern Africa (SFS) to develop a food guide and FBDGs messages and visual illustrations for Tanzania (LOA number SFSD/007/2019). This project formed part of the FOA's larger TCP/SFS/3604 objectives to improve diets and nutrition in Southern Africa by developing and testing food guides and FBDGs for various countries and ultimately to contribute to eradicating of hunger, malnutrition and food insecurity.

Institutional Review Board Statement: The study was conducted according to the guidelines of the Declaration of Helsinki and approved by the Health Research Ethics Committee (HREC) of Stellenbosch University (Ref nr: N19/05/067; 18 June 2019). Furthermore, the research was also approved by the National Institute for Medical Research, Republic of Tanzania (NIMR/HQ/R.8a/Vol.IX/3227; 11 October 2019) and the Zanzibar Health Research Institute (ZAHRI; Ref nr: ZAHREC/02/JULY/2019/09; 31 July 2019).

Informed Consent Statement: Informed consent was obtained from all subjects involved in the study.

Data Availability Statement: Data supporting reported results are available to view on reasonable request.

Acknowledgments: The authors would like to thank all the participants who made data collection of this project possible. We also want to thank the TWG for their invaluable input and contributions to the research. Zanzibar: Ministry of Agriculture, Irrigation, Natural Resources and Livestock (MAINL)—Department of Food Security and Nutrition (Mansura M. Kassim, Ali Mohammed Omar, Ahmed Gharib Khamis, Anisa Kassim Suleiman, Hidaya Ali Abdalla, and Selme Masoud Seif); Ministry of Health, Social Welfare, Elderly, Gender and Children (MoHSWEGC)—Nutrition Unit, Department of Non-Communicable Diseases (NCDs) and Department of Social Welfare and Elderly (Asha A. Salmin, Fatma Ally Said, Omar Abdalla Ali and Sheikha Mohamed Ramia); Zanzibar Food and Drug Agency (ZFDA) (Khadija Ali Shekha); Ministry of Education and Vocational Training (MoEVT)- Department of Health Promotion (Yussuf H. Omar) Higher Learning Institution—State

University of Zanzibar (SUZA) (Omar F. Choum and Jamila Kingwaba Hassan). Milele Zanzibar Foundation (Shemsa N. Msellem); Save the Children (Asha Bilal Pira); Zanzibar NCDs Alliance (Mbaraka Pandu Ali and Mauwa Ame Ameir) and Zanzibar Social Work Association (ZASWA)(Rabia Fadhil Ahmed). Habiba Suleiman, Hidaya Ali Abdalla, Anisa Kassim Suleiman and Selme Masoud Seif, Mtoro Khamis Johari, Hamad Suleiman and Muhiddin Ali Ngwali. Mainland: Waziri Ndonde (FIAT/PAAT); Julius Ntwenya (UDOM); Stella Kimambo (FAO); Grace Moshi (MoHCDGEC); Gibson Kagaruki (NIMR); Mary Mayige (NIMR); Joyce Kinabo (SUA); Peter Mamiro (Late) (SUA); Akwilina W. Mwanri (SUA); Kissa Kulwa (SUA); Theresia Jumbe (SUA); Margareth Natai (Ministry of Agriculture); Ruth Kuandika (Ministry of Agriculture); Andrew Swai (TDA/TANCDA); Fatma Abdallah (TFNC); Hamza Mwangombale (TFNC); Jackson Monela (TFNC); Julieth J. Shine (TFNC); Maria Ngilisho (TFNC); Adeline Munuo (TFNC); Nyamizi Julius Ngassa (TFNC); Aika Lekey (TFNC); Wessy J. Meghji (TFNC); Rehema Mzimbiri (TFNC); AbelaTwinomujuni (TFNC); Belinda Liana (COUNSENUTH); Maria Samlongo (JKCI). *Asante sana.*

Conflicts of Interest: The authors declare no conflict of interest.

References

1. Joint FAO/WHO Consultation on Preparation and Use of Food-Based Dietary Guidelines (1995: Nicosia, Cyprus) & World Health Organization (1998). Preparation and Use of Food-Based Dietary Guidelines/Report of a Joint FAO/WHO Consultation. World Health Organization. Available online: http://apps.who.int/iris/bitstream/handle/10665/42051/WHO_TRS_880.pdf;jsessionid=BA4ECB9959ADB4E25ADE9A8BC7F9911D?sequence=1 (accessed on 22 June 2022).
2. Food and Agriculture Organisation of the United Nations. Food-Based Dietary Guidelines. Background | Food-Based Dietary Guidelines | Food and Agriculture Organization of the United Nations. Available online: https://www.fao.org/nutrition/education/food-dietary-guidelines/background/en/ (accessed on 22 June 2022).
3. Wingrove, K.; Lawrence, M.A.; McNaughton, S.A. Dietary patterns, foods and nutrients: A descriptive analysis of the systematic reviews conducted to inform the Australian Dietary Guidelines. *Nutr. Res. Rev.* **2020**, *34*, 117–124. [CrossRef] [PubMed]
4. Food and Agriculture Organisation of the United Nations. Food Guides. Food-Based Dietary Guidelines. Food Guides | Food-Based Dietary Guidelines | Food and Agriculture Organization of the United Nations. Available online: https://www.fao.org/nutrition/education/food-dietary-guidelines/background/food-guide/en (accessed on 22 June 2022).
5. Food and Agriculture Organisation of the United Nations. FAO's Work in Dietary Guidelines. Food-Based Dietary Guidelines. FAO's Work in Dietary Guidelines | Food-Based Dietary Guidelines | Food and Agriculture Organization of the United Nations. Available online: https://www.fao.org/nutrition/education/food-dietary-guidelines/background/fao-work-dietary-guidelines/en/ (accessed on 22 June 2022).
6. Africa Facts. Facts about Tanzania. Facts About Tanzania | Africa Facts. Available online: https://africa-facts.org/ (accessed on 22 June 2022).
7. Robinson, D.O.; Gaertner, M.; Papageorgiou, C. Tanzania: Growth acceleration and increased public spending with macroeconomic stability. *Yes Afr. Can* **2011**, *21*. Available online: https://openknowledge.worldbank.org/bitstream/handle/10986/2335/634310PUB0Yes0061512B09780821387450.pdf?sequence=1&i#page=37 (accessed on 22 June 2022).
8. World Bank. United Republic of Tanzania Mainland Poverty Assessment. 2015. Microsoft Word-Tanzania Poverty Assessment_Final.docx. Available online: https://www.tralac.org/ (accessed on 22 June 2022).
9. World Bank. Tanzania Mainland Poverty Assessment Report—Executive Summary. 2019. Tanzania-Mainland Poverty Assessment 2019: Executive Summary. Available online: https://www.worldbank.org/en/home (accessed on 22 June 2022).
10. Mayén, A.L.; de Mestral, C.; Zamora, G.; Paccaud, F.; Marques-Vidal, P.; Bovet, P.; Stringhini, S. Interventions promoting healthy eating as a tool for reducing social inequalities in diet in low-and middle-income countries: A systematic review. *Int. J. Equity Health* **2016**, *15*, 1–10. [CrossRef] [PubMed]
11. National Bureau of Statistics (NBS) [Tanzania] and ICF Macro. Tanzania Demographic and Health Survey 2010. Dar es Salaam, Tanzania: NBS and ICF Macro. 2011. Tanzania Demographic and Health Survey 2010 [FR243]. Available online: https://dhsprogram.com/pubs/pdf/FR243/FR243%5B24June2011%5D.pdf (accessed on 22 June 2022).
12. Ministry of Health, Community Development, Gender, Elderly and Children (MoHCDGEC) [Tanzania Mainland]; Ministry of Health (MoH) [Zanzibar]; National Bureau of Statistics (NBS); Office of the Chief Government Statistician (OCGS); ICF. Tanzania Demographic and Health Survey and Malaria Indicator Survey (TDHS-MIS) 2015-16. Dar es Salaam, Tanzania, and Rockville, Maryland, USA: MoHCDGEC, MoH, NBS, OCGS, and ICF. 2016. Tanzania Demographic and Health Survey and Malaria Indicator Survey 2015–2016-Final Report [FR321]. Available online: https://dhsprogram.com/pubs/pdf/fr321/fr321.pdf (accessed on 22 June 2022).
13. Ministry of Health, Community Development, Gender, Elderly and Children (MoHCDGEC); Ministry of Health (MoH); Tanzania Food and Nutrition Centre (TFNC); National Bureau of Statistics (NBS); Office of the Chief Government Statistician (OCGS); UNICEF. *Tanzania National Nutrition Survey Using SMART Methodology (TNNS)*; MoHCDGEC: Mainland, Tanzania; MoH: Zanzibar, Tanzania; TFNC: Dar es Salaam, Tanzania; NBS: Dodoma, Tanzania; OCGS: Zanzibar, Tanzania; UNICEF: Dar es Salaam, Tanzania, 2018.

14. *Global Health Estimates 2020: Deaths by Cause, Age, Sex, by Country and by Region, 2000–2019*; Global Health Estimates: Leading Causes of Death; World Health Organization: Geneva, Switzerland, 2020; Available online: https://www.who.int/ (accessed on 22 June 2022).
15. Mayige, M.; Kagaruki, G. *Tanzania STEPS Survey Report*; National Institute of Medical Research: Dar es Salaam, Tanzania, 2013.
16. Zeba, A.N.; Delisle, H.F.; Renier, G. Dietary patterns and physical inactivity, two contributing factors to the double burden of malnutrition among adults in Burkina Faso, West Africa. *J. Nutr. Sci.* **2014**, *3*, e50. [CrossRef] [PubMed]
17. Popkin, B.M. Global nutrition dynamics: The world is shifting rapidly toward a diet linked with noncommunicable diseases. *Am. J. Clin. Nutr.* **2006**, *84*, 289–298. [CrossRef]
18. Naicker, A.; Venter, C.S.; MacIntyre, U.E.; Ellis, S. Dietary quality and patterns and non-communicable disease risk of an Indian community in KwaZulu-Natal, South Africa. *J. Health Popul. Nutr.* **2015**, *33*, 1–9. [CrossRef]
19. Maletnlema, T.N. A Tanzanian perspective on the nutrition transition and its implications for health. *Public Health Nutr.* **2002**, *5*, 163–168. [CrossRef]
20. Research and Analysis Working Group, United Republic of Tanzania. Brief 4: An Analysis of Household Income and Expenditure in Tanzania. In *Poverty and Human Development Report 2009*; The Povert Eradication and Empowerment Division, Ministry of Finance and Economic Affairs: Dar es Salaam, Tanzania, 2009.
21. World Bank. *Action Plan for the Provision of Vitamins and Minerals to the Tanzanian Population through the Enrichment of Staple Foods*; World Bank: Washington, DC, USA, 2012.
22. Beal, T.; Massiot, E.; Arsenault, J.E.; Smith, M.R.; Hijmans, R.J. Global trends in dietary micronutrient supplies and estimated prevalence of inadequate intakes. *PLoS ONE* **2017**, *12*, e0175554. [CrossRef]
23. Romieu, I.; Dossus, L.; Barquera, S.; Blottière, H.M.; Franks, P.W.; Gunter, M.; Hwalla, N.; Hursting, S.D.; Leitzmann, M.; Margetts, B.; et al. Energy balance and obesity: What are the main drivers? *Cancer Causes Control* **2017**, *28*, 247–258. [CrossRef]
24. Lachat, C.; Otchere, S.; Roberfroid, D.; Abdulai, A.; Seret, F.M.A.; Milesevic, J.; Xuereb, G.; Candeias, V.; Kolsteren, P. Diet and physical activity for the prevention of noncommunicable diseases in low-and middle-income countries: A systematic policy review. *PLoS Med.* **2013**, *10*, e1001465. [CrossRef]
25. Delobelle, P.; Sanders, D.; Puoane, T.; Freudenberg, N. Reducing the role of the food, tobacco, and alcohol industries in noncommunicable disease risk in South Africa. *Health Educ. Behav.* **2016**, *43*, 70S–81S. [CrossRef] [PubMed]
26. United States Department of State. 2018 Report on International Religious Freedom: Tanzania. 2019. Tanzania 2018 International Religious Freedom Report. Available online: https://www.state.gov/ (accessed on 22 June 2022).
27. Du Plessis, L.M.; Daniels, L.C.; Koornhof, H.E.; Samuels, S.; Möller, I.; Röhrs, S. Overview of field-testing of the revised, draft Paediatric Food-Based Dietary Guidelines amongst mothers/caregivers of children aged 0–5 years in the Western Cape and Mpumalanga, South Africa. *S. Afr. J. Clin. Nutr.* **2021**, *34*, 123–131. [CrossRef]
28. Hoddinott, J.; Maluccio, J.; Behrman, J.R.; Martorell, R.; Melgar, P.; Quisumbing, A.R.; Ramirez-Zea, M.; Stein, A.D.; Yount, K.M. The consequences of early childhood growth failure over the life course. In *IFPRI Discussion Paper 01073*; International Food Policy Research Institute: Washington, DC, USA, 2011.
29. Black, R.E.; Victora, C.G.; Walker, S.P.; Bhutta, Z.A.; Christian, P.; De Onis, M.; Ezzati, M.; Grantham-McGregor, S.; Katz, J.; Martorell, R.; et al. Maternal and child undernutrition and overweight in low-income and middle-income countries. *Lancet* **2013**, *382*, 427–451. [CrossRef]
30. Willett, W.; Rockström, J.; Loken, B.; Springmann, M.; Lang, T.; Vermeulen, S.; Garnett, T.; Tilman, D.; DeClerck, F.; Wood, A.; et al. Food in the Anthropocene: The EAT–Lancet Commission on healthy diets from sustainable food systems. *Lancet* **2019**, *393*, 447–492. [CrossRef]
31. Sullivan, G.M.; Sargeant, J. Qualities of qualitative research: Part I. *J. Grad. Med. Educ.* **2011**, *3*, 449–452. [CrossRef]
32. Skinner, D. Qualitative methodology: An introduction. In *Epidemiology: A Research Manual for South Africa*, 2nd ed.; Joubert, G., Ehrlich, R., Eds.; Oxford University Press: Cape Town, South Africa, 2007; pp. 318–327.
33. Love, P.V. Developing and Assessing the Appropriateness of the Preliminary Food-Based Dietary Guidelines for South Africans. Ph.D. Thesis, University of KwaZulu-Natal, Pietermaritzburg, South Africa, 2002. Available online: https://researchspace.ukzn.ac.za/xmlui/handle/10413/9422 (accessed on 22 June 2022).
34. Micha, R.; Mannar, V.; Afshin, A.; Allemandi, L.; Baker, P.; Battersby, J.; Bhutta, Z.; Chen, K.; Corvalan, C.; Di Cesare, M.; et al. *2020 Global Nutrition Report: Action on Equity to End Malnutrition*; Development Initiatives: Bristol, UK, 2020.
35. Hoddinott, J.; Ahmed, A.; Karachiwalla, N.I.; Roy, S. Nutrition behaviour change communication causes sustained effects on IYCN knowledge in two cluster-randomized trials in Bangladesh. *Matern. Child Nutr.* **2018**, *14*, e12498. [CrossRef]
36. Herforth, A.; Arimond, M.; Alvarez-Sanchez, C.; Coates, J.; Christianson, K.; Muehlhoff, E. A Global Review of Food-Based Dietary Guidelines. *Adv. Nutr.* **2019**, *10*, 590–605. [CrossRef]
37. Dhandevi, P.E.M.; Jeewon, R. Fruit and Vegetable Intake: Benefits and Progress of Nutrition Education Interventions—Narrative Review Article. *Iran. J. Public Health* **2015**, *44*, 1309–1321.
38. Mgongo, M.; Hashim, T.H.; Uriyo, J.G.; Damian, D.J.; Stray-Pedersen, B.; Msuya, S.E. Determinants of Exclusive Breastfeeding in Kilimanjaro Region, Tanzania. *Sci. J. Public Health* **2014**, *2*, 631–635. [CrossRef]
39. Dukuzumuremyi, J.P.C.; Acheampong, K.; Abesig, J.; Luo, J. Knowledge, attitude, and practice of exclusive breastfeeding among mothers in East Africa: A systematic review. *Int. Breastfeed. J.* **2020**, *15*, 70. [CrossRef]

40. Du Plessis, L.M.; Daniels, L.C.; Koornhof, H.E.; Solomon, Z.L.; Loftus, M.; Babajee, L.C.; Ronquest, C.; Kleingeld, B.; Greener, C.M.; Burn, K.J. Field-testing of the revised, draft South African Paediatric Food-Based Dietary Guidelines amongst mothers/caregivers of children aged 0–12 months in the Western Cape province, South Africa. *S. Afr. J. Clin. Nutr.* **2021**, *34*, 132–138. [CrossRef]
41. Hoyland, A.; Dye, L.; Lawton, C.L. A systematic review of the effect of breakfast on the cognitive performance of children and adolescents. *Nutr. Res. Rev.* **2009**, *22*, 220–243. [CrossRef] [PubMed]
42. Food and Agriculture Organization (FAO). *Nutrition Guidelines and Standards for School Meals: A Report from 33 Low and Middle-Income Countries*; FAO: Rome, Italy, 2019.
43. Bundy, D.A.P.; de Silva, N.; Horton, S.; Jamison, D.T.; Patton, G.C. *Re-Imagining School Feeding: A High-Return Investment in Human Capital and Local Economies*; License: Creative Commons Attribution CC BY 3.0 IGO; Re-Imagining School Feeding: A High-Return Investment in Human Capital and Local Economies | Save the Children's Resource Centre; World Bank: Washington, DC, USA, 2018.
44. Leme, A.C.B.; Hou, S.; Fisberg, R.M.; Fisberg, M.; Haines, J. Adherence to Food-Based Dietary Guidelines: A Systemic Review of High-Income and Low- and Middle-Income Countries. *Nutrients* **2021**, *13*, 1038. [CrossRef] [PubMed]

 *nutrients*

Article

Nutritional Status, Dietary Intake and Dietary Diversity of Landfill Waste Pickers

Elizabeth C. Swart [1,*], Maria van der Merwe [2], Joy Williams [1], Frederick Blaauw [3], Jacoba M. M. Viljoen [4] and Catherina J. Schenck [5]

[1] Department of Dietetics and Nutrition, Faculty of Community and Health Sciences, University of the Western Cape, Bellville, Cape Town 7530, South Africa; joycyster08@gmail.com
[2] School of Public Health, University of the Western Cape, Bellville, Cape Town 7530, South Africa; smvandermerwe@gmail.com
[3] School of Economic Sciences, North-West University, Potchefstroom Campus, Private Bag X6001, Potchefstroom 2520, South Africa; derick.blaauw@nwu.ac.za
[4] School of Economics and Econometrics, University of Johannesburg, Johannesburg 2092, South Africa; kotiev@uj.ac.za
[5] Department of Social Work, DSI/NRF/CSIR Chair in Waste and Society, University of Western Cape, Bellville, Cape Town 7530, South Africa; cschenck@uwc.ac.za
* Correspondence: rswart@uwc.ac.za

Abstract: The purpose of this study was to investigate and describe the nutritional status, dietary intake and dietary diversity of waste pickers in South Africa, a socioeconomically vulnerable group who makes a significant contribution to planetary health through salvaging recyclable material from dumpsites. Participants were weighed and measured to calculate body mass index (BMI). Dietary intake was recorded using a standardised multipass 24 h recall. Individual dietary diversity scores were derived from the dietary recall data. Data were collected from nine purposefully selected landfill sites located in six rural towns and three cities in four of the nine provinces in South Africa, providing nutritional status information on 386 participants and dietary intake on 358 participants after data cleaning and coding. The mean BMI of the study sample was 23.22 kg/m^2. Underweight was more prevalent among males (22.52%) whilst 56.1% of the females were overweight or obese. The average individual dietary diversity score was 2.46, with 50% scoring 2 or less. Dietary intake patterns were characterised as monotonous, starch-based and lacking vegetables and fruits. The nutritional status, dietary intake and dietary diversity of waste pickers reflect their precarious economic status, highlighting the need for health, social and economic policies to improve access and affordability of nutritious food.

Keywords: waste pickers; nutritional status; dietary intake; dietary diversity; South Africa

Citation: Swart, E.C.; van der Merwe, M.; Williams, J.; Blaauw, F.; Viljoen, J.M.M.; Schenck, C.J. Nutritional Status, Dietary Intake and Dietary Diversity of Landfill Waste Pickers. *Nutrients* **2022**, *14*, 1172. https://doi.org/10.3390/nu14061172

Academic Editors: Penny Love and Colin Bell

Received: 24 December 2021
Accepted: 28 February 2022
Published: 10 March 2022

Publisher's Note: MDPI stays neutral with regard to jurisdictional claims in published maps and institutional affiliations.

1. Introduction

Nutrition is a critical component of health and development, with adequate nutrition aiding the prevention of malnutrition in all its forms, improving productivity and creating opportunities to gradually break the cycles of hunger and poverty.

South Africa is an upper-middle-income country with a culturally diverse population estimated at 59.6 million people [1]. Social, health and economic disparities remain and are aggravated by persistent high poverty, entrenched structural inequality and unemployment [2–4]. The country is experiencing rapid epidemiological transition, with noncommunicable diseases (NCDs) increasing to be the major cause of death, with an estimated 269,000 NCD related deaths annually [5]. This increase as well as the substantial higher burden is particularly noted among lower socioeconomic groups and obese persons [6–8].

The nutrition situation in South Africa is complex and typical of a country in nutrition transition. Undernutrition, notably stunting and micronutrient deficiencies, coexist with a rising incidence of overweight and obesity and the associated consequences such as NCDs [9]. The South African food system is highly commercialised with the majority of households purchasing all their food. The food system is furthermore characterised by cultural and socioeconomic diversity and high levels of income inequality, rendering vulnerable population groups at risk of food insecurity and hunger [10].

The percentage of the population that experienced hunger decreased from 29.3% in 2002 to 11.3% in 2018 and the percentage of people with limited general access to food decreased from 29.1% to 23.8% from 2010 to 2018 [10]. However, the COVID-19 pandemic foregrounded the failures of the food system to provide sufficient, healthy, nutritious food and to serve the most vulnerable people in South Africa, potentially reversing progress achieved in the reduction of hunger of the past two decades. The National Income Dynamics Survey, Coronavirus Rapid Mobile Survey (NIDS-CRAM), a monthly nationally representative panel survey conducted since May 2020, indicated immediate adverse effects on employment and food security and widening inequality. Between March and June 2020, 40% of the NIDS-CRAM sample reported the loss of employment as a result of COVID-19 and 22% of adults and 15% of children were reported to have gone to bed hungry [11,12].

Waste is an unavoidable byproduct of human activities and if accumulated in large quantities, it can lead to degradation of the environment, natural resources and health problems [13]. It is estimated that South Africa generated approximately 54.2 million tons of waste in 2017, of which an estimated 20.7 million tons were recycled. In addition, South Africa produces approximately 31 million tons of food annually, of which an estimated 10 million tons (34.3%) is lost to waste [14].

The commercialisation of waste management, coupled with an increase in recycling practices and unabated unemployment resulted in the survivalist activity of informal waste picking by the poorest and marginalised on landfill sites [15,16]. Their diversification of livelihood is an important strategy towards poverty alleviation. Waste pickers collect, sort, recycle, repurpose and/or sell materials thrown away by others; salvaging and revaluing waste through the recycling value chain [17]. While some rummage in search of necessities, others collect and sell recyclables to middlemen or businesses or work in recycling warehouses or plants owned by their cooperatives or associations [15,18]. In South Africa, there are an estimated 60,000 to 90,000 informal waste pickers operating at different levels [18].

These informal waste pickers operate at the lowest level in the hierarchy of entities that collect and dispose of waste, including municipal waste collection services and recycling structures. Yet they significantly contribute to the economy in general and the recycling economy in particular [19]. Nationally, it is estimated that waste pickers save municipalities up to R750 million per year by collecting waste at no cost and saving landfill space [20]. While waste pickers are typically scorned and treated as a nuisance, their valuable contribution towards waste management and recycling industries are increasingly being recognised and valued by the South African government and stakeholders in the waste industry; with several attempts to integrate waste pickers into formal waste and recycling structures [19]. Their contribution to the economy is in stark contrast to the meagre, and extremely variable, mean financial benefit from informal waste picking trade activities of R451.90 (less than 30 USD) per week in a case study by Schenck et al. in 2018 [21]. Considering that each waste picker in the above case study had a mean of four financial dependents, their household income would seldom exceed the poverty line of R1 268 (82 USD) per person per month [22].

Waste pickers were particularly hard hit by the initial national lockdown regulations following the outbreak of COVID-19, demonstrating their vulnerability to socioeconomic factors. As an informal sector, waste picking was not regarded as an essential service and waste pickers were dismissed for the role they play in the waste sector, resulting in income loss, hunger and reliance on food relief efforts. Reports regarding a lack of consultation and communication with stakeholders demonstrate how the impact of the pandemic extended beyond the immediate risk of infection to further adverse effects related to policy decisions [23].

Although it has been documented that waste pickers obtain food through their activities, the exact contribution of their "findings" to their dietary intake is difficult to quantify, and largely unknown. No scientifically analysed information is available on the nutritional status and dietary intake of waste pickers in South Africa. The objectives of the study were to (1) assess the nutritional status of people who make a living of informal waste picking from landfill sites, (2) to assess the dietary intake of waster pickers, and (3) to assess the dietary diversity of waste pickers.

2. Materials and Methods

This study on the nutritional status, dietary intake and dietary diversity of waste pickers on landfill sites in South Africa, collected quantitative cross-sectional data from nine purposefully selected landfill sites (in six rural towns and three cities and representing both outsourced and municipally managed landfill sites) in four of the nine provinces in South Africa in 2015.

Trained fieldworkers, proficient in the languages spoken on the respective landfill sites, collected socioeconomic data using a questionnaire with quantitative and open-ended explanatory, qualitative questions where more information was required. Dietary intake was recorded by the trained field workers, using a standardised 24 h recall record form and a standardised dietary intake toolkit to assist with quantification [24]. Where possible, interview days were selected to ensure that 24 h recalls represented week and weekend days proportionally. Measurements of weight and height were obtained by one researcher, to limit interobserver variability. Bodyweight was recorded to the nearest 0.1 kg (A&D Personal Precision Scale, Tokyo, Japan) and height to the nearest 1mm (using a portable stadiometer), applying standardised methodology and were used to calculate body mass index (BMI) according to World Health Organization (WHO) criteria [25].

All waste pickers present on the landfill site on the day of data collection who were willing to participate were interviewed. A total of 409 adult waste pickers were interviewed and missing values were recorded for some participants. Discrepancies between numbers of waste pickers with dietary intake, anthropometric measurements and general sociodemographic information are the result of waste pickers' choice not to continue due to time constraints when new trucks with fresh waste arrived. No missing values were recorded for the two smallest landfill sites. Statistical analyses of selected variables in each data set revealed no systematic bias for those participants with missing values (independent *t*-test all *p*-values were ≥ 0.05; paired sample *t*-test after replacing missing values with series mean for continuous variables could not be performed as the standard error of the difference was 0) See Table S1. Subsequently, information on all participants was included in the analyses using a pairwise exclusion analysis by analysis when applicable.

The study was approved by the Senate Research Ethics Committee of the University of the Western Cape (reference number 15/4/24). After receiving information on the process and purpose of the research, participants completed written consent to confirm voluntary participation. Illiterate participants gave verbal consent in the presence of a third person as a witness. Consent forms and information sheets were translated in the most spoken languages and explained in the language best understood. Participants were compensated for their involvement in the form of food worth 1 USD, following the interviews and anthropometric measurements. Questionnaires and data collection forms were anonymised and identified by a code for data analysis purposes. All information acquired was stored securely. Hard copies were stored in a sealed box in a locked cabinet and electronic data was stored on a password-protected computer.

Data Analysis

Anonymised data, recorded in Microsoft Excel 2016, was exported and analysed in IBM SPSS Statistics 25, 2017 for descriptive analysis. Pairwise exclusion was performed and cases were excluded if the variables under investigation were missing.

Anthropometric data was recoded and analysed according to BMI categories. The anthropometric status of male and female waste pickers was compared to the findings of the South African Demographic Health Survey (SADHS) of 2016 reported in their key findings report [26]. The 2016 SADHS used a stratified, two-stage sampling design to provide estimates that were representative of national, provincial and locality type (urban and nonurban areas) and included a sample of 11,083 households. Both studies used the same BMI categories for underweight, overweight and obesity and excluded pregnant women and women who have given birth in the last two months.

The dietary intake data collection was done according to the Centre of Excellence in Food Security Manual on Dietary intake assessment—24 h recall [24]. Quantified 24 h dietary recall data were coded and converted to nutrient intakes using the electronic South African Medical Research Council Food Composition Database. Acceptable macronutrient distribution ranges (AMDR) were used to assess macronutrient intake of participants [27]. Accordingly, the AMDR for carbohydrates is 45–65% of energy, for protein it is 10–35% of energy and for fat it is 20–35% of energy.

Micronutrient intake of participants were compared to the dietary reference intake (DRI) values for men and women aged $\geq$ 19 years [28]. The estimated average requirements (EARs) according to gender and age groups were applied, when established, and adequate intakes (AIs) when EARs were not established.

Dietary adequacy was determined through individual dietary diversity scores. The 24 h recall data was cleaned and coded according to each of sixteen food groups, namely: cereals; white roots and tubers; vitamin A-rich vegetables and tubers; dark green leafy vegetables; other vegetables; vitamin A-rich fruits; other fruits; organ meat; flesh meats; eggs; fish and seafood; legumes, nuts and seeds; milk and milk products; oils and fats; sweets; and spices, condiments and beverages. A dietary diversity score was calculated by summing the number of nine aggregated food groups (starchy staples; dark green leafy vegetables; vitamin A-rich fruits and vegetables; other fruits and vegetables; organ meat; meat and fish; eggs; legumes, nuts and seeds; and milk and milk products) from which foods had been consumed, with a potential score range of 0 to 9 [29]. Only foods consumed in quantities above 1 tablespoon (15 g) were included and each group was only counted once.

3. Results

3.1. Sociodemographic Characteristics

The majority (234) of the 409 participants were male, with six of the sites being male dominated, one site allowed only male waste pickers and the remaining two sites had predominantly female waste pickers. While some of the participants did not know their age, the mean age of the 88.3% of the respondents who knew their age was 39 years. Female participants were on average 5 years older than their male counterparts. The majority (82.1%) of the study participants were black Africans and the remainder was of mixed ancestry (Coloured South Africans). A small proportion (7.9%) of the participants completed secondary education and none had any tertiary education. The income of waste pickers varied greatly depending on the types of waste delivered to landfill sites. This is described elsewhere [30,31]

3.2. Nutritional Status

Of the 409 participants, anthropometric measurements were taken for 386 individuals. The mean BMI of males and females were 20.9 $\pm$ 2.9 kg/m^2 and 26.4 $\pm$ 6.3 kg/m^2 respectively, with most males (74.9%) having a normal BMI, compared to 42.4% of the females. While underweight was more prevalent among males (16.4%) than females (6.7%) (see

Table S2), 50.9% of the females were overweight or obese compared to 8.7% of the males (Table 1).

Table 1. Comparison of BMI of waste pickers with 2016 SADHS* findings.

		Underweight			Overweight/Obese		
	Sex	Severely Thin ≤17	Mildly Thin (>17<18.5)	Total Thin (≤18.5)	Total Overweight (≥25)	Overweight (>25<30)	Obese (≥30)
Waste pickers 2015/16 (Chi-square *p* < 0.001)	Male (*n* = 110)	10 (4.6%)	26 (11.9%)	36 (16.5%)	19 (8.7%)	16 (7.3%)	3 (1.4%)
	Female (*n* = 190)	3 (1.8%)	8 (4.8%)	11 (6.7%)	84 (50.9%)	38 (23.0%)	46 (27.9%)
	TOTAL (*n* = 300)	13 (3.4%)	34 (8.9%)	47 (12.2%)	103 (26.9%)	54 (14.1%)	49 (12.8%)
SADHS 2016 *							
Lowest wealth quantile #	Male	2.9%	7.0%	9.9%	17.4%	14.1%	3.3%
Total		2.1%	7.4%	9.5%	31.3%	20.3%	11.0%
Lowest wealth quantile #	Female	0.6%	2.4%	3.0%	57.3%	27.8%	29.5%
Total		0.5%	2.1%	2.6%	67.7%	26.6%	41.0%

* South African Demographic and Health Survey, 2016 [26]. # The wealth quantile was based on a wealth index that scores households on the number and kinds of consumer goods owned using principal component analyses. Lowest wealth quantile represents 20% of SADHS 2016 [26] population that has fewest assets including those of lowest value.

3.3. Dietary Intake

3.3.1. Energy Intake

The mean daily energy intake, further described in Table 2, was 5664.1 kJ for female participants and 7408.3 kJ for males. These intakes were below the DRI and met 75% and 80% of the daily requirements respectively. The daily energy intake reported by participants varied from 110.4 kJ to 26,147 kJ.

3.3.2. Macronutrient Intake and Distribution

The proportional adequacy of macronutrient and fibre intake is summarised in Table 2. The mean carbohydrate intake was 248.6 g/day for males and 210.6 g/day for females, contributing 191% and 162% of the minimum recommended intake respectively. The mean carbohydrate intake of the majority (51.6%) of participants contributed more than 65% of the total daily energy intake. A higher percentage of women (63.2%) consumed carbohydrates at levels above the acceptable macronutrient distribution range (AMDR), compared to 42.7% of men. Overall, the daily carbohydrate intake varied from 6.0 g to 961.9 g per participant. Carbohydrates were mostly consumed as foods from the cereal and white roots and tubers groups, while 69.5% of the participants also consumed foods from the sweets group (Tables S4 and S5) which includes sugar, sugar-sweetened beverages, sweetened fruit juice and sugary foods such as candies, cookies and cake. The mean intake of added sugar was 40.7 g overall and 44.1 g and 36.0 g for males and females respectively with 42.2% consuming more than 25 g per day. More men (26.9%) had an excessive intake of added sugar compared to women (15.3%). Overall, 38.5% of the participants consumed sugar-sweetened beverages, with intake ranging from 2 to 7 L per person amongst those individuals, while only 0.6% consumed fruit juice. A small proportion of the participants (7.2%) reported having consumed alcohol in the recall period, which contributed to their total carbohydrate intake.

Table 2. Mean intakes of energy, macronutrients, added sugar and total fibre.

Gender		Energy (kJ)	Total Protein (g)	Total Fat (g)	Total Carbohydrate (g)	Added Sugar (g)	Total Fibre (g)
Male (*n* = 205)	Mean	7408.3	61.6	44.85	248.5	44.1	17.7
	SD	4651.8	39.8	38.8	155.5	63.9	12.6
	Median	6809.1	54.6	34.2	233.9	24.0	15.8
	Min	110.4	0.18	0	6.0	0	0
	Max	26,147.0	203.8	225.6	961.9	445.8	92.3
Female (*n* = 153)	Mean	5664.1	43.5	33.8	210.6	36.0	16.4
	SD	3583.8	28.8	29.7	132.4	52.3	12.0
	Median	4917.2	39.12	27.2	184.0	18.0	13.0
	Min	213.6	0	0	10.4	0	0
	Max	23,637.1	154.0	165.0	707.1	330.0	67.0
Total (*n* = 358)	Mean	6662.9	53.8	40.0	232.3	40.7	17.2
	SD	4310.5	36.6	35.5	147.1	59.3	12.3
	Median	6017.2	46.5	29.2	208.9	20.4	14.7
	Min	110.4	0	0	6.0	0	0
	Max	26,147.0	203.8	225.6	961.9	445.8	92.3
Proportional adequacy			Total protein *	Total fat *	Total Carbohydrate *		Fibre *
Inadequate (%)			23.4	50.0	1.0		100
Adequate (%)			76.0	36.6	47.5		0
Excessive (%)			0.6	13.4	51.6		0

* Compared to acceptable macronutrient distribution ranges for adults [27].

The mean total protein intake for women was 43.5 g/day and met 93% of the DRI protein requirement. In addition, the mean total protein intake for men met 109% of the requirement, at 61.5 g/day. However, 23.5% of the participants reported protein intake below the AMDR (Table 2). The total daily protein consumption varied from 0g to 203.8 g per participant for the recalled day. Overall, 61.5% of the participants consumed flesh meat during the 24 h recall period, 8% consumed fish or seafood, 5.5% consumed legumes, seeds and nuts and 4.4% consumed organ meats. Processed meats were consumed by 14.4% of the participants (Table S6).

Very few (16%) of the participants reported fat intake at levels exceeding the recommended macronutrient distribution and half of the study sample (50%) consumed fat at levels below the AMDR. Fat intake per person varied from 0 g to 225.6 g for the 24 h recall.

Only half (50%) of the participants consumed oils and fats, used for cooking, during the reporting period. Amongst the 13% of the study sample with an excessive fat intake, this was attributed to the consumption of foods with high fat content, including fried chicken, vetkoek, chicken feet, Russian sausages (the processed meat product), maas (fermented dairy product), peanut butter, beef with fat and processed beef patties.

3.3.3. Micronutrient Intake

The mean intake of micronutrients was mostly above 80% of the DRI for both male and female participants, as summarised in Table 3. However, the intake of calcium, vitamin E and pantothenic acid was below 80% of the DRI; and very low (below 30% of DRI) for potassium, vitamin C and vitamin D. The DRI for iron in females varies according to age, due to increased iron needs during childbearing years. The intake for all females were analysed collectively and not according to their age, but the mean iron intake for women was above the recommended intake for all age groups. The DRIs for biotin was not met for females and males, but the mean intake level of these micronutrients was above 80% of the DRI.

Table 3. Summary of micronutrient consumption, compared with recommended intake.

	Males (*n* = 205)			Females (*n* = 153)		
	DRI *	Mean	% of DRI	DRI *	Mean	% of DRI
Calcium (mg)	800	286.88	35.9	800	257.52	32.2
Iron (mg)	6	14.35	239.2	5–8.1	11.85	148.1–237.0
Magnesium (mg)	330–350	272.52	77.9–82.6	255–265	210.51	79.4–82.6
Phosphorus (mg)	580	842.39	145.2	580	598.21	103.1
Potassium (mg)	4700	1547.17	32.9	4700	1227.84	26.1
Sodium (mg)	1500	1463.90	97.6	1500	883.4	58.9
Manganese (mg)	2.3	2.43	105.7	1.8	2.28	126.7
Zinc (mg)	9.4	11.88	126.4	6.8	8.66	127.4
Vitamin A RE (µg)	625	527.17	84.3	500	503.58	100.7
Thiamin (mg)	1.0	1.62	162.0	0.9	1.43	158.9
Riboflavin (mg)	1.1	1.12	101.8	0.9	0.90	100.0
Niacin (mg)	12	20.69	172.4	11	12.52	113.8
Vitamin B6 (mg)	1.1	3.52	320	1.1	2.27	206.4
Folate (µg)	320	365.67	114.2	320	333.62	104.3
Vitamin B12 (µg)	2.0	4.19	209.5	2.0	2.83	141.5
Pathothenate (mg)	5	3.40	68.0	5	2.24	44.8
Biotin (µg)	30	28.26	94.2	30	25.92	86.4
Vitamin C (mg)	75	17.77	23.7	60	20.18	33.6
Vitamin D (µg)	10	2.47	24.7	10	1.49	14.9
Vitamin E (mg)	12	6.39	53.3	12	6.43	53.6

* Estimated adequate requirement (EAR), where established, or adequate intake (AI) [28].

The mean sodium intake was 58.9% of the DRI for women and 97.6% of the DRI for males. The salt intake of 16% of the study sample, who reported excessive sodium intake, was checked and included farm-style sausages and Russian sausages (processed meat products), maas (fermented dairy product) and sugary drinks. No added sodium intake was reported in the 24 h recall data and sodium was therefore mostly consumed through processed foods.

3.4. Dietary Diversity

The average individual dietary diversity scores were low and did not vary greatly between sites (Figure 1 and Tables S3–S6). A low dietary diversity score is associated with food insecurity as well as a risk of micronutrient deficiency. Very low consumption of foods from the fruit and vegetable groups, particularly vitamin A-rich fruits and vegetables, was reported. Less than a quarter (23.8%) of the participants consumed milk and milk products, while only 5.5% of the participants consumed legumes, nuts and seeds. The types of food reported to be consumed by waste pickers are summarised in Table 4, according to food groups.

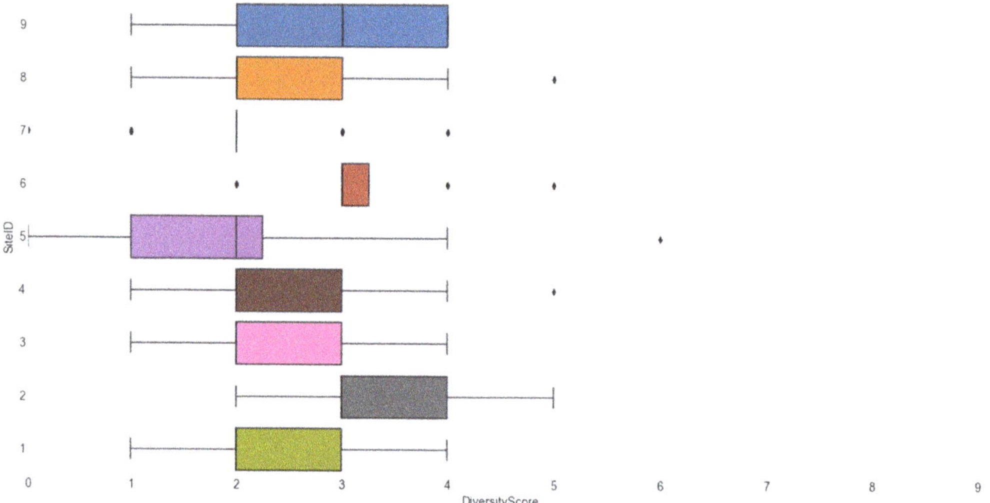

Figure 1. Box plot of dietary diversity scores per site. Sites 1, 3 and 7 were in urban areas. All other sites were in rural areas.

Table 4. Summary of types of food consumed by waste pickers, according to food groups.

Food Groups	Types of Foods Reported to be Consumed by Waste Pickers
Cereals	Bread, breakfast cereal, hot cross bun, maize porridge, oats porridge, pasta, pizza, rice, samp (split white corn), scone, sorghum porridge, vetkoek (fried dough), Weet-Bix, wholegrain breakfast cereal
White roots and tubers	Potato, sweet potato
Vitamin A rich vegetables and tubers	Carrot, pumpkin
Vitamin A rich fruits	None
Dark green leafy vegetables	Amaranth leaves, broccoli, cabbage, Swiss chard
Other vegetables	Beetroot, garlic, lettuce, mixed vegetables, onion, sweet pepper, tomato
Other fruits	Apple, banana, grape, lemon, pear
Organ meat	Beef liver, chicken feet, chicken giblets, chicken head, sheep liver
Flesh meats	Bacon, beef, beef patty, beef sausage, chicken, mutton, ostrich, polony, pork, salami, turkey, vienna sausage
Fish and seafood	Fish, pilchards
Eggs	Egg
Legumes, nuts and seeds	Beans, peanuts, soya mince
Milk and milk products	Cheese, cheese spread, milk, yogurt
Oils and fats	Canola oil, margarine, mayonnaise, nondairy creamer, peanut butter, sunflower oil
Sweets	Carbonated cold drink, chocolate, chocolate coated bar, cold drink squash, condensed milk, dairy fruit juice mix, jam, glucose drink, sweets, sweetened orange juice, sugar
Spices, condiments and beverages	BBQ sauce, beer, Bovril (meat extract paste), coffee, curry sauce, fruit chutney, gravy, instant soup, mango achar, Rooibos tea, sorghum beer, soup powder, spirits, tomato sauce, tea, vinegar, wine

4. Discussion

While dietary intake of the study participants varied greatly, the total energy consumption was below the recommended intake range. With regards to macronutrient distribution, most participants had an excessive consumption of carbohydrates, adequate intake of proteins and inadequate intake of fats. None of the participants met the recommended intake for fibre. Micronutrient intake was mostly adequate according to the recommended dietary intake, except for potassium, vitamin C and vitamin D. While the sodium intake for the study population was at an acceptable level, a proportion of the population had a very high sodium intake due to the type of foods found at the sites where they worked. The overall dietary diversity among study participants was very low, with little variation between sites.

Consumption of a variety of foods is required to ensure adequate nutrient intake. The availability and affordability of highly processed foods are considered important drivers of poor nutrition [32]. On the other hand, dietary patterns characterised by higher intakes of unprocessed foods are linked to more positive health outcomes [33]. However, for the most vulnerable groups, nutrient-rich foods such as animal-source foods, fruits and vegetables are not affordable [34,35].

The quarterly labour force survey of the fourth quarter of 2020 reports the official unemployment rate in South Africa at 32.5% [36]. Many individuals and households, therefore, rely on social support as a source of income, with 31% and 44.3% respectively being dependent on social grants in 2018 [10]. Thirty-four per cent of participants benefitted from social grants with either themselves or other household members receiving a grant [30].

Waste picking is not a choice or preference, but rather a source of income for those desperate to make a living, and in some instances also a source of food. People with a low income may consume a less healthy diet due to energy-dense foods being relatively cheap sources of energy, but with a low nutrient density. A healthy diet remains unaffordable for most South Africans [34,35]. Waste pickers are regarded as vulnerable to economic instability. A high level of food insecurity—not having physical and/or economic access to sufficient food to meet dietary needs for a productive and healthy life at all times—was measured among the waste pickers included in this study and reported elsewhere [21]. Overall, 20% of the participants reported going to bed hungry at night and a further 18% reported going for a whole day and night without eating. In this study, two participants reported not eating or drinking anything during the previous day. Food sources included food brought from home, bought as ready meals, found on the landfill site or shared by other waste pickers [21].

The morbidity and mortality rates related to obesity and NCDs are higher among people from socioeconomically disadvantaged groups, due to poor food choices which mostly include energy-dense foods and low nutrition density [37]. The prevalence of obesity among the adult population in South Africa is increasing alarmingly, with more than a quarter of the female adult population estimated to be overweight and almost a third obese [38]. The prevalence of overweight and obesity in women has increased by 21% between 1998 and 2016 [9,38]. In addition, 31% of adult men are overweight or obese, an increase of 2% since 1998 [9,38]. In contrast to the burden of overweight and obesity, 3% of the South African female adult population and 10% of the male adult population are underweight and a dual burden of child undernutrition and adult obesity exists with 27% of children under the age of 5 years being stunted—a sign of chronic undernutrition [38].

The BMI of more than half of the waste pickers in this study was within the normal range. Underweight and overweight/obesity of female and male waste pickers were statistically significantly different. Of concern, the prevalence of underweight among male participants was double that of the national male population [38]. As illustrated in Table 1, underweight in both males and females were higher than the lowest wealth quantile in the national SADHS, 2016 [25]. Overweight and obesity was low in males, but in the female study population it was comparable to the lowest wealth income quantile of the SADHS 2016 albeit lower compared to the general South African population [38]. It is known that the prevalence of overweight and obesity varies among different population groups and the severity of obesity increases with increasing wealth [38], although overweight and obesity is very high in South African females regardless of wealth quantile. The lower prevalence of overweight and obesity among the study population, compared to the national population is therefore potentially attributed to their socioeconomic status. The prevalence of overweight and obesity however remains a concern, due to the association with NCDs, the leading cause of death in South Africa [2]. Access to public health services for NCD related treatment for waste pickers may be limited especially given the high opportunity cost as their daily survival is based on their engagement at the landfill sites at critical times of waste deposits.

Representing the leading cause of mortality worldwide (71% of all deaths), the prevalence of NCD multimorbidity is exploding in low- and middle-income countries, and substantially increasing in sub-Saharan Africa, particularly among lower socioeconomic groups and obese persons [9–11]. A large study (n = 1025) amongst waste pickers in the largest dumpsite in Latin America found 32.6% to be overweight and 21.1% obese, with a 24.2% prevalence of hypertension 24.2% and 10.1% of diabetes [39]. Another study from Brazil established that 29.5% of 253 male-dominant (86.2%) pushcart waste pickers in the city of Santos were overweight or obese [40]. Auler et al. (2014) reported a much higher prevalence of overweight (51.1%) and obesity (25.7%) with hypertension (32.8%) and diabetes (11.4%) among 268 waste pickers in southern Brazil [41].

In the absence of national data on the dietary intake of adult South Africans, a recent review of dietary surveys in the adult South African population from 2000 to 2015 [42] concluded that energy intake varied from low intakes in informal settlements to very high intakes in urban centres. Macronutrient intake varied similarly to energy intakes but remain within acceptable minimum distribution ranges [42]. Overall, food consumption patterns in the country have dramatically shifted towards a concerning overall increase in daily energy intake, a diet of sugar-sweetened beverages, an increase in the proportion of processed foods and animal source foods, added sugar and a shift away from vegetables [43].

Many low-income households consume monotonous, low-quality diets, typically cereal-based and lacking in vegetables, fruit and animal-source foods [44]. Monotonous diets and a lack of dietary diversity is closely associated with food insecurity [45]. The risk of micronutrient deficiencies also increases when dietary diversity is low [46] although it should be noted that mandatory fortification of maize meal, wheat flour and bread [47] is assumed to have contributed to the micronutrient content of diets of the waste pickers. Dietary diversity is a proxy for nutrient adequacy [29]. National food diversity was evaluated in 2009 as a proxy for food security, with a dietary diversity score below 4 regarded to reflect poor dietary diversity and poor food security [48]. Overall, a national level dietary diversity score of 4.02 reflects that the majority of South Africans consumed a diet low in dietary variety, albeit with significant provincial differences. Low dietary diversity correlated with socioeconomic status, lower-income groups having the lowest dietary diversity score of 2.93 [48]. At 2.465, the average dietary diversity score of waste pickers included in this study is lower than the average score of low-income groups in the national evaluation; implying that waste pickers are at explicit risk of food insecurity and micronutrient deficiencies due to inadequate diets.

Food groups most commonly consumed, according to the evaluation of national dietary diversity [48], included starchy staples, meat and fish, dairy and vegetables other than vitamin A rich food. Eggs, legumes and vitamin A-rich fruits and vegetable were the least consumed among the national population. Similarly, the most commonly deficient food groups observed in a recent review of dietary surveys in South Africa included vegetables and fruits, and milk and milk products [42]. Based on reported intake over the past 24 h, 49% and 59% of participants in the 2016 Demographic and Health Survey reported consuming fruit and vegetables respectively, with variations between population groups and according to income status [38]. It is a global trend for low-income households to spend less on fresh produce than higher-income households [49]. Of further concern, the consumption of fresh vegetables in South Africa declined whilst consumption of ultra-processed foods increased dramatically between 1994 and 2012 [43].

Fruit and vegetable consumption amongst waste pickers in this study is much lower than the reported national consumption and confirms that these foods are not accessible and/or affordable to socially vulnerable populations. A low intake of fruits and vegetables are of particular concern in relation to micronutrient deficiencies. Meat was the food most commonly collected on landfill sites [21]. This would explain why most of the waste pickers consumed foods from the meat and fish group, contrary to what would be expected due to their socioeconomic status as it was sometimes obtained from the sites and not bought with household income. The variation in overall dietary intake among waste pickers in this study is likely an indication of varied access to food retrieved from the landfill site.

Frequent consumption of food products not beneficial to health, such as sugary drinks, is of concern. According to the 2016 South African Demographic Health Survey [38], 36% of participants reported drinking any sugar-sweetened beverage (607 mL on average) and 14% reported drinking fruit juice (304 mL on average). A similar proportion of waste pickers in this study reported consuming sweetened beverages. The absence of access to water on the landfill sites [30] may be contributing to the consumption of sugary beverages.

With food production, processing and marketing being driven by profit, ultraprocessed food is becoming increasingly available and affordable [50]. At the same time, food security and the nutritional status of the most vulnerable population groups are likely to deteriorate further due to the health and socioeconomic impacts of the COVID-19 pandemic [44]. In South Africa, it is estimated that 9.34 million people faced high levels of acute food insecurity by the end of 2020 with projections of a further increase with regards to the number of people affected.

Particular attention should be paid, through policy interventions, to curb increasing food prices and the energy cost of food preparation. Access to and the affordability of nutrient-dense foods, by the most vulnerable and marginalised populations, including waste pickers as a particularly vulnerable group, should be improved. This may include, but are not limited to, policies to make healthy food options more affordable with consideration of a subsidy for the vulnerable, increased targeted social protection including food relief, and an extension of the health promotion levy to tax other unhealthy food items.

4.1. Significance for Public Health

Waste pickers are an extremely vulnerable population from an occupational health perspective [31]. In addition, their precarious income generation and fierce relative competition for "spots" on landfill sites to seek out a livelihood compromise their health seeking behaviour, yet they contribute to reduction of waste to landfill sites and environmental sustainability. Knowledge and understanding of the lived realities of waste pickers should guide service delivery planning by community nutrition and public health practitioners. Targeted service delivery to this vulnerable population could include the provision of health screening, primary health care and chronic care through mobile outreach services. In addition, a coordinated approach to extend service delivery to include aspects of social services will go a long way to improve their lives. These extended services could include applying for identity documents to allow the opportunity to access social grants, and registration on the indigent programme for free basic services such as water and electricity. The rendering of services to this population should be planned and executed with sensitivity and understanding of their unique working (and often also living) conditions. Waste pickers cannot afford the opportunity cost of not working for a day.

4.2. Limitations

Reported dietary intake is subject to bias caused by systematic underreporting, over-reporting or omission of foods by an individual during the interviewing process. The convenience sample approach, including all participants available on the day and willing to participate, may introduce sample selection bias. It was not feasible to repeat the dietary intake assessment as the informal and fluid nature of the setting made it difficult to locate the respondents. The findings from this study are not regarded as representative of other settings.

Supplementary Materials: The following supporting information can be downloaded at: https://www.mdpi.com/article/10.3390/nu14061172/s1, Table S1: Missing values per landfill site; Table S2: Body mass index of participants; Table S3: Body mass index classification according to landfill site; Table S4: Summary of food consumption per site by food group; Table S5: Consumption of food and beverages that are not part of a healthy eating plan; Table S6: Average individual dietary diversity score per site.

Author Contributions: E.C.S., C.J.S., F.B. and J.M.M.V.; conceptualised the study. E.C.S., J.W. and M.v.d.M.; prepared the first draft. All authors have read and agreed to the published version of the manuscript.

Funding: This research/study was made possible thanks to funding by the DSI-NRF Centre of Excellence in Food Security (UID 91490), South Africa Project number 140504.

Institutional Review Board Statement: The Senate Research Ethics Committee of the University of the Western Cape provided ethics approval for the study (reference number 15/4/24).

Informed Consent Statement: The information sheet advised potential participants that a report or article may be written on the findings of the research, and that the authors will ensure that their identity will not be disclosed in such event. After receiving information on the process and purpose of the research, participants completed written consent to confirm voluntary participation. Illiterate participants gave verbal consent in the presence of a third person as a witness. Consent forms and information sheets were translated in the most spoken languages and explained in the language best understood by the participant. The manuscript does not contain any individual person's data in any form.

Data Availability Statement: The anonymised datasets used and analysed during the current study are available from the corresponding author on reasonable request.

Acknowledgments: We thank Ria Laubscher, South African Medical Research Council, and Eben Gerryts, for their assistance with analyses.

Conflicts of Interest: The funders had no role in the design of the study; in the collection, analyses, or interpretation of data; in the writing of the manuscript, or in the decision to publish the results.

References

1. Statistics South Africa. *Mid-Year Population Estimates 2020*; Statistical Release P0302 [Internet]; Statistics South Africa: Pretoria, South Africa, 2020. Available online: http://www.statssa.gov.za/publications/P0302/P03022020.pdf (accessed on 4 April 2021).
2. Pillay-van Wyk, V.; Msemburi, W.; Laubscher, R.; Dorrington, R.E.; Groenewald, P.; Glass, T.; Bradshaw, D. Mortality trends and differentials in South Africa from 1997 to 2012: Second National Burden of Disease Study. *Lancet Glob. Health* **2016**, *4*, e642–e653. [CrossRef]
3. World Bank. *Overcoming Poverty and Inequality in South Africa—An Assessment of Drivers, Constraints and Opportunities Overcoming Poverty and Inequality in South Africa*; World Bank: Washington, DC, USA, 2018.
4. Ataguba, J.E.; Akazili, J.; McIntyre, D. Socioeconomic-related health inequality in South Africa: Evidence from General Household Surveys. *Int. J. Equity Health* **2011**, *10*, 48. [CrossRef]
5. World Health Organization. *Noncommunicable Diseases Progress Monitor 2020*; World Health Organization: Geneva, Switzerland, 2020.
6. Bigna, J.J.; Noubiap, J.J. *The Rising Burden of Non-Communicable Diseases in Sub-Saharan Africa*; The Lancet Global Health; Elsevier Ltd.: London, UK, 2019; Volume 7, pp. e1295–e1296.
7. Gouda, H.N.; Charlson, F.; Sorsdahl, K.; Ahmadzada, S.; Ferrari, A.J.; Erskine, H.; Leung, J.; Santamauro, D.; Lund, C.; Aminde, L.N.; et al. Burden of non-communicable diseases in sub-Saharan Africa, 1990–2017: Results from the Global Burden of Disease Study 2017. *Lancet Glob. Health* **2019**, *7*, e1375–e1387. [CrossRef]
8. Statistics South Africa. *Mortality and Causes of Death in South Africa, 2016: Findings from Death Notification*; Statistics South Africa: Pretoria, South Africa, 2016.
9. Shisana, O.; Labadarios, D.; Rehle, T.; Simbayi, L.; Zuma, K.; Dhansay, A.; Reddy, P.; Parker, W.; Hoosain, E.; Naidoo, P.; et al. *The South African National Health and Nutrition Examination Survey, 2012 (SANHANES-1)*; HSRC Press: Cape Town, South Africa, 2014.
10. Statistics South Africa. *General Household Survey 2018*; Statistics South Africa: Pretoria, South Africa, 2018.

11. Jain, R.; Budlender, J.; Zizzamia, R. *The Labour Market and Poverty Impacts of COVID-19 in South Africa*; SALDRU Working Paper No. 264; SALDRU, UCT: Cape Town, South Africa, 2020.

12. Van der Berg, S.; Zuze, L.; Bridgman, G. *The Impact of the Coronavirus and Lockdown on Children's Welfare in South Africa Evidence from NIDS-CRAM Wave 1*; NIDS-CRAM Policy Brief 11; Stellenbosch University, Department of Economics: Stellenbosch, South Africa, 2020.

13. Mote, B.N.; Kadam, S.B.; Kalaskar, S.K.; Thakare, B.S.; Adhav, A.S.; Muthuvel, T. Occupational and environmental health hazards (physical & mental) among rag-pickers in Mumbai slums: A cross-sectional study. *Sci. J. Public Health* **2016**, *4*, 1–10.

14. Oelofse, S.; Polasi, T.; Haywood, L.; Musvoto, C. *Increasing Reliable, Scientific Data and Information on Food Losses and Waste in South Africa*; CSIR: Pretoria, South Africa, 2021; Available online: https://wasteroadmap.co.za/wp-content/uploads/2021/06/17-CSIR-Final_Technical-report_Food-waste.pdf (accessed on 19 September 2021).

15. Waste Pickers | WIEGO [Internet]. Women in Informal Employment: Globalizing and Organizing. Available online: https://www.wiego.org/informal-economy/occupational-groups/waste-pickers (accessed on 12 April 2021).

16. Qotole, M.; Xali, M.; Barchiesi, F. The Commercialisation of Waste Management in South Africa [Internet]. Occasional. McDonald, D.A., Bond, P., Eds.; Municipal Services Project. 2001. Available online: www.queensu.ca/msp (accessed on 12 April 2021).

17. Schenck, R.; Blaauw, P.F. The work and lives of street waste pickers in Pretoria—A case study of recycling in South Africa's urban informal economy. *Urban Forum.* **2011**, *22*, 411–430. [CrossRef]

18. Godfrey, L.; Oelofse, S. Historical review of waste management and recycling in South Africa. *Resources* **2017**, *6*, 57. [CrossRef]

19. Samson, M.; Timm, M.; Chidzungu, S.; Dladla, T.; Kadyamadare, N.; Maema, G. *Lessons from Waste Picker Integration Initiatives: Development of Evidence Based Guidelines to Integrate Waste Pickers into South African Municipal Waste Management Systems*; Final Technical Report: Johannesburg Case Study; DEFF and DST: Pretoria, South Africa, 2020.

20. Farber, T. Meet the Bin Scavengers Saving SA R750-Million a Year [Internet]. *Times Live.* 2016. Available online: https://www.timeslive.co.za/sunday-times/news/2016-05-08-meet-the-bin-scavengers-saving-sa-r750-million-a-year/ (accessed on 13 April 2021).

21. Schenck, C.J.; Blaauw, P.F.; Viljoen, J.M.; Swart, E.C. Social work and food security: Case study on the nutritional capabilities of the landfill waste pickers in South Africa. *Int. Soc. Work* **2018**, *61*, 571–586. [CrossRef]

22. Statistics South Africa. *National Poverty Lines 2020*; Statistical Release P0310.1; Statistics South Afric: Pretoria, South Africa, 2020.

23. Harrisberg, K. How Are We Meant to Eat? A South African Waste Picker on Life under Lockdown [Internet]. *Thomson Reuters Foundation News*, 2020. Available online: https://news.trust.org/item/20200407102057-bcmya/?fbclid=IwAR1zi_Lj-KoMrgbhGt3FgKKenVUqJ2sWkl6GtjsxxD6Q3k_qr4LTtjA4OGk (accessed on 7 April 2021).

24. Faber, M.; Kunneke, E.; Wentzel-Viljoen, E.; Wenhold, F. *Dietary Intake Assessment–24 Hour Recall. Centre of Excellence in Food Security*; DST-NRF Centre of Excellence in Food Security/South African Medical Research Council: Cape Town, South Africa, 2016.

25. World Health Organization; Body Mass Index-BMI [Internet]. Available online: http://www.euro.who.int/en/health-topics/disease-prevention/nutrition/a-healthy-lifestyle/body-mass-index-bmi (accessed on 17 May 2021).

26. Statistics South Africa. *South African Demographic and Health Survey 2016: Key Indicators Report*; Statistics South Africa: Pretoria, South Africa, 2017.

27. Institute of Medicine. *Dietary Reference Intakes for Energy, Carbohydrates, Fiber, Fat, Fatty Adds, Cholesterol, Protein, and Amino Acids*; National Academies Press: Washington, DC, USA, 2005.

28. Dietary Reference Intakes (DRIs): Recommended Dietary Allowances and Adequate Intakes, Vitamins. Food and Nutrition Board, Institute of Medicine, National Academies [Internet]. Available online: https://www.ncbi.nlm.nih.gov/books/NBK56068/table/summarytables.t2/?report=objectonly (accessed on 21 June 2021).

29. Food and Agriculture Organization of the United Nations. *Guidelines for Measuring Household and Individual Dietary Diversity*; Kennedy, G., Ballard, T., Dop, M., Eds.; Food and Agriculture Organization of the United Nations: Rome, Italy, 2013.

30. Schenck, C.J.; Blaauw, P.F.; Swart, E.C.; Viljoen, J.M.M.; Mudavanhu, N. The management of South Africa's landfills and waste pickers on them: Impacting lives and livelihoods. *Dev. S. Afr.* **2019**, *36*, 80–98. [CrossRef]

31. Schenck, C.J.; Blaauw, P.F.; Viljoen, J.M.M.; Swart, E.C. Exploring the potential health risks faced by waste pickers on landfills in South Africa: A socio ecological perspective. *Int. J. Environ. Res. Public Health* **2019**, *16*, 2059. [CrossRef] [PubMed]

32. Popkin, B.M.; Adair, L.S.; Ng, S.W. The global nutrition transition and the pandemic of obesity in developing countries. *Nutr. Rev.* **2012**, *70*, 3–21. [CrossRef]

33. Wrottesley, S.V.; Pisa, P.T.; Norris, S.A. The influence of maternal dietary patterns on body mass index and gestational weight gain in urban black South African women. *Nutrients* **2017**, *9*, 732. [CrossRef] [PubMed]

34. Temple, N.J.; Steyn, N.P. The cost of a healthy diet: A South African perspective. *Nutrition* **2011**, *27*, 505–508. [CrossRef]

35. Vermeulen, H. A Balanced Food Basked Approach to Monitor Food Affordability in South Africa. Ph.D. Nutrition. 2020. Available online: https://repository.up.ac.za/bitstream/handle/2263/73175/Vermeulen_Balanced_2020.pdf?sequence=1&isAllowed=y (accessed on 19 September 2021).

36. Statistics South Africa. *Quarterly Labour Force Survey. Quarter 4: 2020*; Statistical Release P0211; Statistics South Africa: Pretoria, South Africa, 2021.

37. Moreira, P.A.; Padra, P.D. Educational and economic determinants of food intake in Portuguese adults: A cross-sectional survey. *BMC Public Health* **2004**, *4*, 58. [CrossRef] [PubMed]

38. South African National Department of Health. *South Africa Demographic Health Survey 2016*; Department of Health, South Africa; Statistics South Africa; South African Medical Research Council: Pretoria, South Africa, 2019.
39. Resende Nogueira Cruvinel, V.; Pintas Marques, C.; Cardoso, V.; Rita Carvalho Garbi Novaes, M.; Navegantes Araújo, W.; Angulo-Tuesta, A.; Escalda, P.M.F.; Galato, D.; Brito, P.; da Silva, E.N.; et al. Health conditions and occupational risks in a novel group: Waste pickers in the largest open garbage dump in Latin America. *BMC Public Health* **2019**, *19*, 581.
40. Rozman, M.A.; de Ezevedo, C.H.; de Jesus, R.R.C.; Filho, R.M.; Junior, V.P. Anemia in recyclable waste pickers using human driven pushcarts in the city of Santos, southeastern Brazil. *Braz. J. Epidemiol.* **2010**, *13*, 326–336.
41. Auler, F.; Nakashima, A.T.A.; Cuman, R.K.N. Health conditions of recyclable waste pickers. *J. Community Health* **2014**, *39*, 17–22. [CrossRef] [PubMed]
42. Mchiza, Z.J.; Steyn, N.P.; Hill, I.; Kruger, A.; Schönfeldt, H.; Nel, J.; Wentzel-Viljoen, E. A review of dietary surveys in the adult South African population from 2000 to 2015. *Nutrients* **2015**, *7*, 8227–8250. [CrossRef]
43. Ronquest-Ross, L.C.; Vink, N.; Sigge, G.O. Food consumption changes in South Africa since 1994. *S. Afr. J. Sci.* **2015**, *111*, 12. [CrossRef]
44. FAO; IFAD; UNICEF; WFP; WHO. *The State of Food Security and Nutrition in the World 2020. Transforming Food Systems for Affordable Healthy Diets*; FAO: Rome, Italy, 2020.
45. Faber, M.; Schwabe, C.; Drimie, S. Dietary diversity in relation to other household food security indicators. *Int. J. Food Saf. Nutr. Public Health* **2009**, *2*, 1–15. [CrossRef]
46. Chakona, G.; Shackleton, C. Minimum Dietary Diversity Scores for Women Indicate Micronutrient Adequacy and Food Insecurity Status in South African Towns. *Nutrients* **2017**, *9*, 812. [CrossRef]
47. Friesen, V.M.; Mbuya, M.N.N.; Aaron, G.J.; Pachón, H.; Adegoke, O.; Noor, R.A.; Swart, R.; Kaaya, A.; Wieringa, F.T.; Neufeld, L.M. Fortified foods are major contributors to intakes of vitamin A and iodine, but not iron, in diets of women of reproductive age in four African countries. *J. Nutr.* **2020**, *150*, 2183–2190. [CrossRef] [PubMed]
48. Labadarios, D.; Steyn, N.P.; Nel, J. How diverse is the diet of adult South Africans? *Nutr. J.* **2011**, *10*, 33. [CrossRef]
49. Institute of Medicine, National Research Council. *Supplemental Nutrition Assistance Program: Examining the Evidence to Define Benefit Adequacy*; Caswell, J., Yaktine, A., Eds.; National Academies Press: Washington, DC, USA, 2013.
50. Sanders, D.; Joubert, L.; Greenberg, S.; Hutton, B.; Bautista Hernández, F.A.; Rojas, I.C.D. *At the Bottom of the Food Chain: Small Operators versus Multinational Corporations in the Food Systems of Brazil, Mexico and South Africa*; Economic Justice Network: Cape Town, South Africa, 2018.

Systematic Review

Dietary Characteristics and Influencing Factors on Chinese Immigrants in Canada and the United States: A Scoping Review

Ping Zou [1,*], Dong Ba [2], Yan Luo [3], Yeqin Yang [4], Chunmei Zhang [4], Hui Zhang [5] and Yao Wang [6,*]

[1] School of Nursing, Nipissing University, 222 St. Patrick Street, Suite 618, Toronto, ON M5T 1V4, Canada
[2] Faculty of Health Sciences, McMaster University, 1280 Main Street West, Hamilton, ON L8S 4L8, Canada; bad@mcmaster.ca
[3] Faculty of Nursing, Health Science Center, Xi'an Jiaotong University, No. 76 Yanta West Road, Xi'an 710061, China; luoyan0904@xjtu.edu.cn
[4] School of Nursing, Wenzhou Medical University, Chashan Higher Education Park, Wenzhou 325035, China; yangyq@wmu.edu.cn (Y.Y.); sallyzcm@wmu.edu.cn (C.Z.)
[5] Department of Cardiology, Guizhou Provincial People's Hospital, Guiyang 550002, China; zhanghui88640@163.com
[6] Xiang Ya School of Nursing, Central South University, Changsha 410013, China
* Correspondence: pingz@nipissingu.ca (P.Z.); yao.wang@csu.edu.cn (Y.W.); Tel.: +416-642-7003 (P.Z.)

Citation: Zou, P.; Ba, D.; Luo, Y.; Yang, Y.; Zhang, C.; Zhang, H.; Wang, Y. Dietary Characteristics and Influencing Factors on Chinese Immigrants in Canada and the United States: A Scoping Review. *Nutrients* 2022, 14, 2166. https://doi.org/10.3390/nu14102166

Academic Editors: Colin Bell and Penny Love

Received: 18 March 2022
Accepted: 18 May 2022
Published: 23 May 2022

Publisher's Note: MDPI stays neutral with regard to jurisdictional claims in published maps and institutional affiliations.

Abstract: Background: Chinese immigrants are an integral part of Canadian and American society. Chinese immigrants believe diet to be an important aspect of health, and dietary behaviours in this population have been associated with changes in disease risk factors and disease incidence. This review aims to summarize the characteristics of the dietary behaviours of Chinese immigrants and the associated influencing factors to better inform individual, clinical, and policy decisions. Methods: This scoping review was written in accordance with PRISMA guidelines. MEDLINE, PsychINFO, CINAHL, AgeLine, ERIC, ProQuest, Nursing and Allied Health Database, PsychARTICLES, and Sociology Database were utilized for the literature search. Articles were included if they explored dietary or nutritional intake or its influencing factors for Chinese immigrants to Canada or the United States. Results: A total of 51 papers were included in this review. Among Chinese immigrants in Canada and the United States, the intake of fruits and vegetables, milk and alternatives, and fiber were inadequate against national recommendations. Chinese immigrants showed increased total consumption of food across all food groups and adoption of Western food items. Total caloric intake, meat and alternatives intake, and carbohydrate intake increased with acculturation. Individual factors (demographics, individual preferences, and nutritional awareness), familial factors (familial preferences and values, having young children in the family, and household food environment), and community factors (accessibility and cultural conceptualizations of health and eating) influenced dietary behaviours of Chinese immigrants. Discussion and Conclusion: Efforts should be undertaken to increase fruit, vegetable, and fibre consumption in this population. As dietary acculturation is inevitable, efforts must also be undertaken to ensure that healthy Western foods are adopted. It is important for healthcare providers to remain culturally sensitive when providing dietary recommendations. This can be achieved through encouragement of healthy ethnocultural foods and acknowledgement and incorporation of traditional health beliefs and values into Western evidence-based principles where possible.

Keywords: diet; influencing factors; Chinese; immigrants; review; Canada; United States

1. Introduction

Chinese immigrants are an integral part of the functioning North American society. According to the 2016 Canadian census, Chinese-born individuals represented the second largest visible minority group in Canada, with a population exceeding 1.5 million [1]. In the United States, there are nearly 2.5 million Chinese immigrants, with China being the top country of origin for immigrants [2]. Chinese immigrants to North America have been demonstrated to hold significant human capital [3] and have been regarded as a key minority group in society. Chinese immigrants usually maintain their cultural beliefs and values after immigration [4].

Chinese immigrants identify diet as an important factor in health maintenance and disease mitigation and management [5]. Traditional Chinese Medicine emphasizes the nutritional use of foods, herbs, and the concept of humoral balance between hot and cold foods to achieve good, stable health [6,7]. Additionally, the promotion of healthy dietary behaviours in the Chinese immigrant population have been shown to influence cardio-vascular risk factors, including HDL cholesterol, blood pressure, and waist circumference, as well as cardiovascular disease incidence [8]. The impact of dietary factors on health outcomes has been well documented in the existing literature, with dietary and lifestyle behaviours being key to managing a variety of chronic diseases [9].

There is a need to summarize and identify the influencing factors of Chinese immi-grants' dietary behaviours to facilitate the design and implementation of specific dietary interventions in local North American communities. An existing scoping review discussed the nutritional health of Canadian immigrants of diverse ethnicities. They identified a dearth of evidence regarding how acculturation influences dietary habits [10]. This review included but was not specific to Chinese immigrants and missed psychosocial factors other than acculturation. A more focused narrative review on dietary habits of South Asians in Western countries identifies the examination of the intake of food groups as well as the intake of macro- and micro-nutrients as a key future step [11]. Moreover, a literature review of dietary patterns and influencing for Arabic-speaking immigrants and refugees in Western societies highlighted a gap in nutrition research for their respective immigrant groups [12]. Two literature reviews, which focused on Chinese immigrant populations, examined dietary behaviours in the context of specific diseases such as diabetes and focused on cultural influence [13,14]. Thus, there is a need to better synthesize the evidence on the specific dietary practices, changes, and influencing factors of Chinese immigrant groups.

This review strives to summarize the characteristics and influencing factors of di-etary behaviours in the Chinese immigrant populations in Canada and the United States. The ecosocial theory proposes a framework of intertwined factors from micro to macro levels [15]. This framework has been previously applied in qualitative studies assessing influencing factors on dietary behaviours in Chinese immigrants with hypertension [16]. This present review employs the theory to elucidate the influencing factors on dietary behaviours among Chinese immigrants at personal, familial, and community levels. The research questions of this review are: (1) what are characteristics of diet behaviours among Chinese immigrants in Canada and the United States; (2) what are the factors influencing their dietary behaviours and changes?

2. Methods

The protocol and reporting of the results of this scoping review were based on the PRISMA statement [17].

2.1. Eligibility Criteria

Studies were included if they measured dietary or nutritional intake of Chinese immigrant adults to Canada or the United States, or if they investigated influencing factors of dietary behaviours for Chinese immigrant adults to Canada or the United States. We included qualitative, quantitative, and mixed-method studies.

Studies were excluded if they: (a) had results not specific to the population of Chinese adult immigrants to Canada and the United States, (b) did not have data relevant to dietary influencing factors or dietary characteristics, (c) were theses or review papers, or (e) did not have an accessible electronic text document.

2.2. Information Sources

Various health-related, psychological, sociological, and educational science databases, including MEDLINE, PsycINFO, CINAHL, AgeLine, ERIC, ProQuest, Nursing and Allied Health Database, PsycARTICLES, Sociology Database, and Education Research Complete, were selected for the literature search.

2.3. Search Strategy and Selection of Evidence

The databases were systematically searched using a combination of the following keywords: (Chin*) AND (immigr* OR migrant* or migrat*) AND (North America* OR Canad* OR United States OR America* OR New York* OR Toronto* OR California* OR Vancouver* OR Montreal) AND (Diet* OR food* OR eat* OR cook* OR nutri* OR carbohydrate* OR Protein* OR Fat or Fats OR Milk* OR Dairy OR Sugar* OR Potassium* OR Calcium* OR Sodium* OR Vitamin* Or fruit* Or vegetable* OR meat* OR rice* OR grain*)) OR ti((Chin*) AND (immigr* OR migrant* or migrat*) AND (North America* OR Canad* OR United States OR America* OR New York* OR Toronto* OR California* OR Vancouver* OR Montreal) AND (Diet* OR food* OR eat* OR cook* OR nutri* OR carbohydrate* OR Protein* OR Fat or Fats OR Milk* OR Dairy OR Sugar* OR Potassium* OR Calcium* OR Sodium* OR Vitamin* OR fruit* OR vegetable* OR meat* OR rice* OR grain*). The citations were exported into EndNote to remove any duplicates. The titles and abstracts of all citations were screened for relevance based on the established eligibility criteria. All eligible articles were searched for full text documents and the full text documents were carefully reviewed, with reasons for exclusion noted. Furthermore, tables of contents of key journals were hand-searched for the previous 2 years, and the reference lists of all eligible articles were manually searched. The most recent search was conducted in February 2022.

2.4. Quality Assessment

The Critical Appraisal Skills Program Checklists have been used as quality assessment tools for the included articles [18]. These checklists are not designed to generate a final quantitative score. Rather, they draw attention to the elements of a rigorous study and evaluate the study as a whole. Using these checklists, we were able to classify the quality of included papers as low, moderate, or high. Two researchers (DB & PZ) independently evaluated each article; any discrepancies in ratings were discussed along with the guidelines until consensus was reached. Papers rated as low quality were excluded. Thus, all papers included in this review were moderate to high quality.

2.5. Data Extraction

Data were independently extracted by two reviewers (DB & PZ) based on pre-determined criteria. From each article, various data, including authors, year of publication, study population, research design, recruitment method, sample size, sample characteristics, comparison group, features of the intervention, outcomes, measurements, significant findings, limitations, and future direction were extracted. The data were collected and organized into an Excel spreadsheet. The reviewers discussed disagreements in data extraction until consensus was reached.

2.6. Synthesis of Results

Once the data was organized in Excel, descriptive statistics were used to present the characteristics of the included studies. Thematic analysis was then used to summarize the findings of each research question. Categorization results were compared among reviewers (DB & PZ) and any disagreements among reviewers were resolved with a consensus decision. Due to the heterogeneity of the measurement tools used by the included studies, a meta-analysis was not performed since attempting to combine different measurements for the same variable would be inappropriate.

3. Results

3.1. Characteristics of Included Studies

In this review, 51 studies were included (Figure 1). Thirty-six (36/51, 70.6%) papers were original independent studies, and fifteen (15/51, 29.4%) papers were secondary analyses of prior surveys. Five (5/51, 9.8%) independent studies and ten (10/51, 19.6%) secondary studies included populations besides Chinese immigrants. Considering only Chinese immigrant participants, the sample size of independent studies ranged from 10 to 805 participants, and the sample size of secondary studies ranged from 120 to 2061. Six studies (6/51, 11.8%) were qualitative, two (2/51, 3/9%) were mixed-method, and forty-three (43/51, 84.3%) were quantitative. Thirty-three (33/51, 64.7%) were conducted in the United States of America, four (4/51, 7.8%) were conducted on participants in both Canada and America, and fourteen (14/51, 27.4%) were conducted in Canada (Table 1).

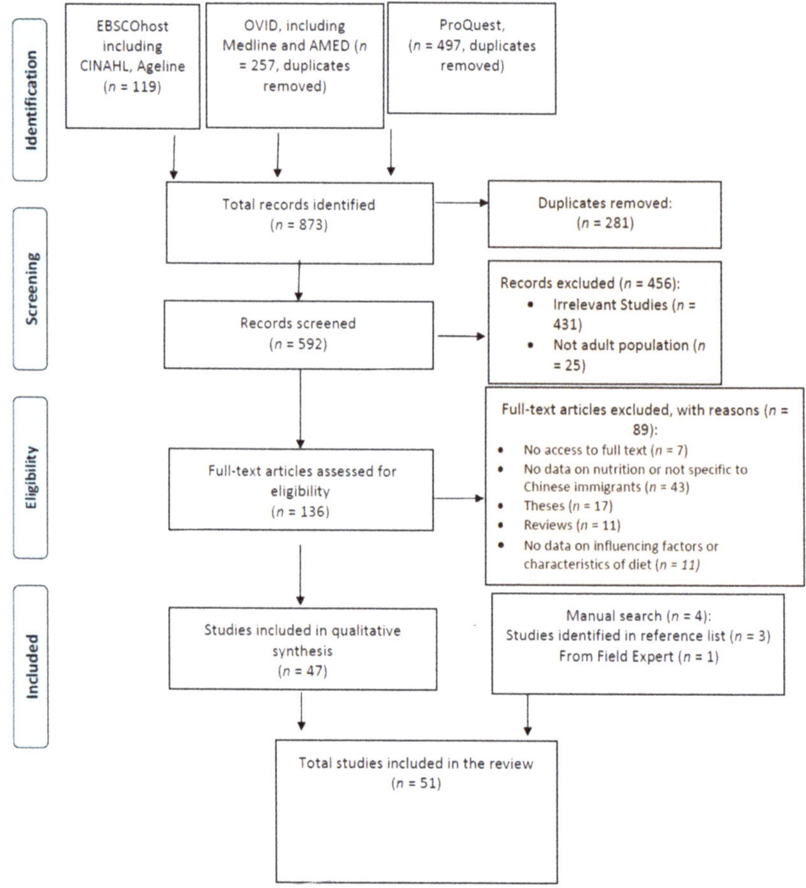

Figure 1. Study selection flow diagram.

Table 1. Characteristics of included studies.

	Author Year	Research Setting	Research Design	Sampling
1	Rodriguez et al. (2020) [19]	USA, community	Quantitative, cross-sectional secondary analysis from the Mediators of Atherosclerosis in South Asians Living in America (MASALA) and the Multi-ethnic study of atherosclerosis (MESA), which were longitudinal studies	3927 + 889 participants free of CVD
2	Beasley et al. (2019) [8]	New York City, USA	Quantitative, cross-sectional survey	1973 Chinese immigrants, as part of Chinese American Cardiovascular Health Assessment
3	Chai et al. (2019) [20]	Delaware, USA, university setting	Quantitative cross-sectional survey	172 Asian students, students, 108 from China or Taiwan
4	Kirshner et al. (2019) [21]	New York City, USA, community	Quantitative, cross-sectional survey	2071 Chinese American New Yorkers, data from the Chinese American Cardiovascular Health Assessment
5	Higginbottom et al. (2018) [7]	Alberta, CAN, community	Quantitative, ethnography and interviews	23 Chinese Canadian perinatal women
6	Zou (2018) [16]	Greater Toronto Area, CAN, telephone	Qualitative, telephone interviews	30 aged Chinese-Canadian participants who received the DASHNa-CC
7	An (2017) [22]	United States, online	Quantitative, cross-sectional survey	505 Chinese living in US, online
8	Liu et al. (2017) [23]	Greater Toronto Area, CAN, community	Mixed methods, qualitative focus group, and quantitative small-scale cross-sectional survey	12 female Chinese immigrants, who had been in Canada for 10 years or less
9	Lu et al. (2017) [24]	Vancouver, Toronto, Halifax and St Catherines, CAN, community	Quantitative, cross-sectional survey	100 Chinese immigrants aged 25+
10	Wang et al. (2017) [23]	California, USA	Quantitative, cross-sectional	2122 Asian adults, 658 of whom were Chinese
11	Zou (2017) [25]	Greater Toronto Area, CAN, classroom and community setting	Quantitative RCT, intervention received DASHNa-CC, control received usual care	61 Chinese Canadians 45+ with hypertension but not on medications
12	Yi et al. (2016) [26]	New York City, USA, community	Quantitative, cross-sectional data obtained from New York City Community Health Survey	555 Chinese America adults with hypertension, 144 South Asian adults with hypertension, 5987 Non Hispanic white adults with hypertension

Table 1. *Cont.*

	Author Year	Research Setting	Research Design	Sampling
13	Tseng et al. (2015) [27]	Philadelphia, USA, community	Quantitative, longitudinal study	312 Chinese immigrant women
14	Corlin et al. (2014) [28]	Boston, USA, community-based	Quantitative, cross-sectional	147 Chines immigrants and 167 US born whites participating in Community assessment of Freeway Exposure and Health study
15	Wyat et al. (2014) [29]	New York City, USA, community	Quantitative longitudinal, surveys over 4 years.	805 Chinese Americans aged 65+, foreign-born
16	Adekunle et al. (2013) [30]	Greater Toronto Area, CAN	Quantitative, cross-sectional survey, predictive factor analysis	250 Chinese Canadian respondents, representing household averaging four people
17	Wong et al. (2013) [31]	New York City, USA, community	Quantitative cross-sectional	125 older (50+) Chinese persons
18	Tam et al. (2012) [32]	Toronto and Vancouver, CAN, community	Quantitative, Cross sectional	1050 postmenopausal Canadian women, 421 of whom were recent Chinese migrants, 216 of whom migrated to the West before age 21
19	Tseng et al. (2012) [33]	Philadelphia, USA, community	Quantitative, Cross-sectional surveys	437 healthy premenopausal Chinese Immigrant women
20	Alonge et al. (2011) [34]	Houston, Texas, USA	Quantitative, Cross-sectional surveys	213 Chinese, Mexican and Nigerian immigrants, 52 of whom were Chinese.
21	Lv et al. (2011) [35]	USA, community	Quantitative, Quasi-experimental study with a nested design and pre- and post design	151 first generation Chinese American mothers between 35 and 55
22	Rosenmoller et al. (2011) [36]	CAN, community	Quantitative, cross-sectional sub-study of the Multi-cultural Community health assessment Trial, study cohort	120 Chinese-born people living in Canada
23	Tam et al. (2011) [37]	Toronto and Vancouver, CAN, community	Quantitative, cross-sectional	1051 postmenopausal Canadian women, 383 of whom were recent Chinese migrants, 156 of whom migrated to the west before age 21
24	Tseng et al. (2011) [38]	Philadelphia, USA, community	Quantitative cross-sectional surveys	436 healthy premenopausal Chinese Immigrant women
25	Liu et al. (2010) [39]	Philadelphia, USA	Quantitative, cross-sectional	243 Chinese Americans who were part of study on diet and breast density
26	Bell et al. (2009) [40]	British Columbia, CAN, professionally facilitated support group	Qualitative, ethnography	96 Chinese Canadian participants in cancer support groups

Table 1. *Cont.*

	Author Year	Research Setting	Research Design	Sampling
27	Chesla et al. (2009) [41]	USA	Qualitative, comparative interview	20 Chinese American couples, one with diabetes
28	Kwok et al. (2009) [42]	Toronto, CAN community	Quantitative, cross-sectional survey	106 Chinese Canadians
29	Osypuk et al. (2009) [43]	Four USA cities, community	Quantitative, secondary analysis from the Multi-ethnic study of atherosclerosis, which was a longitudinal study	1902 Study participants
30	Washington et al. (2009) [5]	California, USA, two Chinese senior care facilities	Qualitative semi-structured interviews	13 participants, aged 65 years or older, who had a diagnosis of type 2 diabetes
31	Hislop (2008) [44]	Vancouver, CAN	Quantitative, cross-sectional	504 Chinese adult immigrants
32	Kandula et al. (2008) [45]	USA: Baltimore, Chicago, Forsyth County, LA, NYC, St Paul	Cross-sectional data from Multi-Ethnic Study of Atherosclerosis	1255 Hispanics and 737 Chinese participants
33	Lu et al. (2008) [46]	Western Canada, community	Qualitative, semi-structured interviews	10 individuals
34	Taylor et al. (2007) [47]	Seattle, USA, community	Quantitative, cross-sectional survey	495 Chinese immigrants
35	Babbar et al. (2006) [48]	New York City, USA, family care center	Mixed Methods, concurrent triangulation of cross-sectional study and qualitative surveys	300 Chinese American Women
36	Fang et al. (2006) [49]	New York City, USA, hospital based	Quantitative, case control study	187 foreign-born Chinese stroke cases and 204 controls matched
37	Walker et al. (2006) [50]	USA, community	Quantitative, development of prognostic model	359 Chinese American women, ambulatory, ages 20–90
38	Liang et al. (2004) [51]	Washington, DC, USA, community	Qualitative, focus groups	54 Chinese American women aged 50+
39	Lv et al. (2004) [52]	Pennsylvania, USA, community	Quantitative, cross-sectional self-administered survey	399 Chinese Americans, 18+ in Pennsylvania
40	Lv et al. (2003) [53]	Pennsylvania, USA	Quantitative, cross-sectional survey	399 Chinese Americans, 18+ in Pennsylvania
41	Kelemen et al. (2003) [54]	Hamilton, CAN, community	Quantitative, development of a tool (involved multiple 24 h recalls, items tabulated, assessed, and included into existing tool)	74 immigrants, 25 of whom were Chinese
42	Satia-Abouta et al. (2002) [55]	Seattle, Vancouver, USA	Quantitative, Secondary analysis of data from Chinese Women's Health Project, cross sectional	244 adult females of Chinese ethnicity

Table 1. *Cont.*

	Author Year	Research Setting	Research Design	Sampling
43	Wu et al. (2002) [56]	Los Angeles County, USA	Quantitative case-control study	523 cases, Asian American women between ages of 25–74 at time of diagnosis of breast cancer were identified through the LA Cancer Surveillance program. 160 were Chinese. 594 controls were selected from the neighbourhood. 228 were Chinese
44	Liou et al. (2001) [57]	New York City, USA, community	Quantitative, cross-sectional survey	600 health Chinese Americans between 25 and 70 years of age
45	Satia et al. (2001) [58]	Seattle, USA and Vancouver, CAN community	Quantitative, Secondary analysis of data from Chinese Women's Health Project, cross-sectional	244 adult females of Chinese ethnicity
46	Satia et al. (2001) [59]	Seattle, USA, and Vancouver, CAN	Quantitative, Secondary analysis of data from Chinese Women's Health Project, including cross-sectional survey, development of a measurement tool	244 adult females of Chinese ethnicity
47	Satia et al. (2000) [60]	Seattle, USA, community	Qualitative interviews and focus groups, qualitative groundwork to develop quantitative dietary survey tool	42 Chinese American women
48	Whittemore et al. (1995) [61]	USA: LA, San Francisco, Hawaii, CAN: Vancouver, Toronto In community and lab	Quantitative case control study	1655 prostate cancer cases were identified through cancer registries in Hawaii, LA, SF, Vancouver, and Ontario Cancer Registry, 283 of whom were Chinese Americans. 1645 controls, 272 of whom were Chinese Americans
49	Choi et al. (1990) [62]	Boston, USA	Quantitative, cross-sectional surveys	346 healthy elderly Chinese aged 60–96
50	Whittemore et al. (1990) [63]	USA: LA, San Francisco, Vancouver CHINA: Hangzhou, Ningbo Hospitals	Quantitative, case-control study	805 Chinese North American patients were identified from the British Columbia Cancer Registry
51	Newman et al. (1982) [64]	New York City, USA, community	Quantitative, cross-sectional surveys	102 Chinese immigrant mothers

3.2. Dietary Characteristics

3.2.1. Food Consumption

Eleven (11/51, 31.5%) studies reported that Chinese immigrants in Canada and the United States regularly consumed fruits or vegetables at an estimated amount of 2.7 to 3.6 servings/day. [5,26,29,30,32,42,44,47,49,53,61]. Studies reported a frequency of vegetable consumption ranging from at least weekly [53] to more than 3–5 times per week [42,49]. Chinese immigrants also self-reported a significant amount of vegetable intake [5,30]. Studies reported a frequency of fruit intake at least weekly [53] to more than 4–5 times weekly [42,49]. Five studies reported the quantity of combined fruit and vegetable consumption per day (3.5 servings per day, 3.6 servings per day, 566.3 g per day, 2.7 fruits or vegetables per day) [26,29,44,47,61]. One study reported fruit and vegetable consumption to be 2.7 and 2.6 servings, respectively, where each serving was equal to half a cup. The estimated consumption of fruits and vegetables among Chinese immigrants thus ranged from 2.7 to 3.6 cups per day [26,32,44,47,61].

Four (4/51, 7.8%) studies reported that grains are an important source of energy for Chinese immigrants in Canada and the United States [5,32,49,53]. Studies reported more than 90% of Chinese immigrants consumed grains more than 5 times a week [49] and, specifically, rice was consumed daily [5,53]. Grains were the main source of carbohydrates for recent Chinese immigrants [32].

Five studies (5/51, 9.8%) reported a low frequency or quantity of dairy consumption in Chinese immigrants in Canada and the United States, ranging from less than one serving per day to 1.6 servings/day [20,31,32,49,53,56]. Two studies found that Chinese immigrants consumed less than 1 serving of dairy a day, one found consumption of 1.6 cups per day, and yet another found that consumption of most dairy products occurred on a monthly basis [20,31,32,53]. One study reported that about 40% of immigrants consumed milk and alternatives less than or equal to 5 times per week [49].

Three (3/51, 5.9%) studies reported that most Chinese immigrants in Canada and the United States regularly consume meat and alternatives, including soybeans, fish, and eggs [32,49,53,56]. More than 75% of Chinese immigrants were reported to consume soybeans over 3 times per week [49]. Consumption of eggs, soy, and fish were reported to be higher in comparison with Caucasians [32,56], and consumption of meat and alternatives in general was reported to occur on a weekly basis [53].

3.2.2. Macronutrient Intake

Five (5/51, 9.8%) studies reported varying daily carbohydrate consumption and inadequate fibre consumption among Chinese immigrants in Canada and the United States [20,32,34,54,62]. Three studies reported carbohydrates making up over 50% of total energy intake, [62] with daily intake ranging from 185 to 258 g per day [34,54,62]. Three studies reported fibre consumption of 12–14 g per day [20,34,54]. Four (4/51, 7.8%) studies reported quantities of protein consumption ranging from 1.1–1.5 g/kg/day, with one study reporting 89 g per day [37,54,62,63]. Chinese immigrants were found to consume more protein than Caucasians, South Asians, and Europeans [37,54]. Meat and fish were the largest source of protein in Chinese immigrants [63]. Seven (7/51, 13.7%) studies reported that fats were consumed regularly by Chinese immigrants in Canada and the United States, that the major source of fat intake was through cooking oils, and that there was a tendency to reduce fat among this population [42,49,54,57,60–62]. The daily quantity of fat consumption was reported to be 67 g per day in the general adult populations, and 54 to 57 g per day in males 60+ years of age, and 29 to 42 g per day in females 60+ years of age [54,62]. Studies reported daily consumption of vegetable oil and consumption of butter, margarine, or lard to be more than three times a week [49,53]. Thirty per cent of daily diet intake was reported to be derived from fats [62], with cooking oil being a major source [60,63], although Chinese immigrants had a tendency to reduce their intake of fat through methods such as limiting fried foods [42,57].

3.2.3. Micronutrient and Caloric Intake

Four (4/51, 7.8%) studies reported a wide range of calcium intake in Chinese immigrants in Canada and the United States, ranging from less than 333 mg/day to 612 mg/day [20,35,48,50]. Three of these studies focused on Chinese women exclusively [35,48,50]. Ten (10/51, 19.6%) studies reported that Chinese immigrants in Canada and the United States had a lower daily caloric intake than Caucasians [8,19,21,28,32,34,54,61,63,65]. Caloric consumption was reported by two studies at 1592 kcal/day and 1736 kcal/day [8,34]. The dietary quality of Chinese immigrants was found to be 66.2/110 on the AHEI score, a scale for which values above 80 indicates a good diet and lower than 50 indicates a poor diet [19].

3.2.4. Dietary Changes since Immigration

Eight (8/51, 15.7%) studies reported an increase in total consumption of food across all food groups and an adoption of Western food items in Chinese immigrants in Canada and the United States [5,20,23,36,46,52,59,64]. Three studies reported an increase in overall consumption of all food groups [23,52,59]. Two studies reported an increase in fruit and vegetable consumption specifically [36,64]. Four studies reported an increase in consumption of meat or dairy specifically [23,36,59,64]. Three studies reported that most Chinese immigrants consumed a traditional diet daily [5,23,46]. However, the frequency of consumption of traditional food is conflicting, with two studies reporting decreased consumption of traditional foods such as rice [52,64], and one reporting increased consumption of traditional foods [20].

Five (5/51, 9.8%) studies examined other dietary changes, such as adopting of food items (e.g., bread rolls, cakes, or pies), snacking between meals, drinking milk, and eating at fast-food restaurants [7,42,46,52,59,64]. Western dietary acculturation was found to be associated with higher-fat dietary behaviour [58,59]. Three studies reported that Chinese immigrants adopted certain Western foods, such as pizza, cereal, bread, pasta, or pies [23,52,59]. Three studies reported snacking between meals [42,59,64]. Three studies reported that breakfast was usually the first meal to be qualitatively altered (i.e., consuming peanut butter on toast) [42,46,59]. Additionally, studies found increased self-reported fast food and "junk food" consumption, the latter presumably referring to high-fat, high-calorie foods with little nutritional value according to a definition provided by the World Health Organisation [7,46,66]. However, one study found an increase in dietary variety and a decrease in high-fat foods since immigration [36].

3.3. Factors Influencing Dietary Behaviours

3.3.1. Acculturation and Its Associations with Diet

Acculturation was reported to be associated with increased total energy intake, carbohydrate intake, and meat intake, although there were conflicting findings with regards to the association between acculturation and overall dietary quality, fruit and vegetable consumption, and fat consumption. Studies reported that an increased degree of acculturation or number of American friends was associated with increased total caloric intake [27,45], increased meat intake frequency, and increased beef, pork, and dairy intake [27,53]. There was no association reported between acculturation and consumption of fish or tofu [27]. Studies reported increased total carbohydrate and sugar intake with increased degree of acculturation [27,45]. The association between dietary quality and acculturation was conflicting. Various studies found acculturation to be associated with no significant effect on diet quality [23], that dietary quality was highest in less-acculturated participants [22], and that more-acculturated subjects had better dietary variety and adequacy [39,45,53]. Conflicting associations were also reported between acculturation and fruit and vegetable consumption [27,30,53,58]. Two studies reported increased fruit and vegetable consumption with acculturation [53,58]. However, one reported decreased expenditure on fruits and vegetables with acculturation, and yet another reported no change associated with acculturation [27,30]. Conflicting findings were reported on the association between acculturation and fat consumption [33,42,53,58,60]. Four studies found an association between acculturation and higher fat dietary behaviour [27,53,58,59]. One study reported an association of acculturation with decreased fat behaviour, and two others reported an association of acculturation with more fat-reducing behaviours and food items [42,45,58].

Five (5/51, 9.8%) studies reported conflicting findings on the effect of length of stay on dietary pattern changes of Chinese immigrants in Canada and the United States [24,28,31,47,53]. Three studies found that length of stay was not associated with changes in dietary patterns [28,31,53]. However, studies also reported that recent immigrants have a higher intake of meat and are more likely to use calcium supplements [31], that intake of fruit and vegetables decreased with length of stay [47], and that associations of a healthy traditional Chinese diet weakened with length of stay [24]. Studies also reported decreased added sugar consumption with length of stay, although there was no association between total sugar consumption and length of stay [20,34].

3.3.2. Individual Factors

Twelve (12/51, 23.5%) studies reported that the dietary behaviours of Chinese immigrants in Canada and the United States were associated with individual demographics, individual preferences, and conceptualisations of health and nutritional awareness [7,16,23–25,33,38,41,42,53,59,60]. Individual demographics such as younger age, higher education, employment, and female gender were found to be associated with dietary acculturation [24,53,59]. Higher education and employment were associated with greater intake of energy and sugar [33]. Individual preferences and values influenced dietary behaviours, such as the desire for continued consumption of traditional foods such as ethno-cultural vegetables and the decision to decrease fat intake [23,42,55,60]. Personal awareness of nutritional information was associated with the adoption of Western dietary practices, and lack of information on healthful diets was a challenge for disease management [25,55]. The safety, quality, and freshness of products were major deciding factors in dietary behaviours and change [7,23,30,53,60]. Regularity, moderation, and the concept of maintaining yin/yang or "hot/cold" balance affected dietary behaviours as they were also seen as important aspects of healthy eating [7,41,42,51,60].

Six (6/51, 11.8%) studies reported associations between one's personal health and life circumstances, such as nature of life experiences and language fluency with North American Chinese immigrants' dietary behaviours [7,16,23,30,38,55]. One's personal health condition also affects dietary behaviours, with conditions like pregnancy or having a disease associated with increased dairy consumption and healthier dietary behaviours, respectively [7,16]. Additionally, positive life events or experiences were associated with greater energy intake and healthier dietary behaviours [16,38]. Stress was associated with lower overall dietary intake, but a greater energy density and percentage of energy from fat [41]. The amount of time one has to prepare traditional and healthy meals is also an influencing factor [7,16,55]. Language was another influencing factor, acting as a barrier to acquiring safe food and underlying the decision to purchase cultural vegetables [23,30].

3.3.3. Familial Factors

Eight (8/51, 15.7%) studies investigated the association of familial factors on dietary behaviours and changes in Chinese immigrants in Canada and the United States, including familial preferences and values, having young children in family, having older relatives and male partners, and household food environment [7,16,23,39,53,55,59,60]. Familial preferences and values have a major influence as individuals may seek family support in dietary decisions, and family preferences can be a facilitator or barrier to healthy eating [7,16,23,55,60]. One study found that older relatives and male partners prefer traditional foods, and two studies found that having young children was associated with the adoption of Western dietary practices [25,53]. Having young children was also associated with increased fruit and vegetable consumption [53]. Respondents in households with more high-fat foods exhibited higher fat-related dietary behaviour [58].

3.3.4. Community Factors

Sixteen (16/51, 31.4%) studies investigated the association of community factors with the dietary behaviours of Chinese immigrants in Canada and the United States, including accessibility, cultural conceptualisations of health and eating, and community programs and people [5,7,16,23,30,33,40–43,51–53,55–57]. Accessibility of traditional foods is influenced by cost, availability, store location, and convenience of access. Chinese immigrants may be influenced in their decision to maintain traditional Chinese meals or eat healthily depending on the convenience of preparing traditional, healthy meals [7,16,39,52,53,55]. Limited access to traditional or healthy foods was reported to be associated with dietary acculturation or unhealthy diet behaviours [7,16,40,52,55]. Cost and socio-economic environment affects both the healthiness and acculturative degree of diets [7,23,53,55,60]. The cultural conceptualisations of health and diet also play an important role in dietary changes and behaviours. Eight studies reported that Chinese immigrants believe that diet is a very important part of health [5,7,30,40–42,51,56]. Specifically, the traditional concept of balancing "hot" vs. "cold" foods, or yin vs. yang foods, influences dietary decisions [7,24,42,51,60]. Finally, community nutrition education workshops and materials were identified as facilitators of healthy eating [16]. Neighbourhoods with higher immigrant populations were associated with healthier consumption and lower high-fat and processed food consumption [43].

4. Discussions

4.1. Summary of Findings

Chinese immigrants regularly consumed fruits or vegetables at an estimated amount of 2.7 to 3.6 servings/day. The consumption of milk and alternatives in Chinese immigrants was reported with values ranging from less than one serving per day to 1.6 servings/day. Grains, particularly rice, were an important source of energy for Chinese immigrants. Fiber consumption was inadequate according to national recommendations. Protein and fat consumption was sufficient. With immigration, Chinese immigrants showed an increased total consumption of food across all food groups and adoption of Western food items. Chinese immigrants also exhibited more snacking between meals and Westernisation of breakfast since immigration, although healthfulness of changes since immigration was variable. Total caloric intake, meat and alternatives intake, and carbohydrate intake increased with acculturation, although there were conflictive results regarding fruit and vegetable consumption and fat consumption. Individual factors associated with dietary behaviours of Chinese immigrants were individual demographics, preferences, and conceptualisations of health and nutritional awareness. Familial factors included familial preferences and values, having young children in family, and household food environment. Community factors included accessibility, cultural conceptualisations of health and eating, community programs and people.

4.2. Dietary Characteristics

The dietary behaviours of the Chinese immigrant population are discussed by comparing them against national dietary guidelines, while healthy eating is defined as dietary behavious that align with the United States or Canadian dietary guidelines. The findings of this review indicated that Chinese immigrants regularly consumed fruits and vegetables at an estimated amount of 2.7 to 3.6 servings/day. This amount is lower than the United States Department of Agriculture (USDA)'s Dietary Guidelines 2020–2025, recommending two cups of fruits and three cups vegetables for someone consuming 2000 calories per day [67]. Although these guidelines distinguish fruits and vegetables as different food groups, most findings from the included studies in this review do not. One study did find fruit and vegetable consumption to be 2.7 and 2.6 servings, respectively, where each serving was equal to half a cup—rendering each serving half of USDA standards. From this study and the comparatively low total consumption from other studies, it can be reasonably hypothesised that requirements from neither of the fruits or vegetables are met.

These findings are consistent with a previous systematic review, which indicated only 0.5% to 20% of Asian American adults to be reaching the recommended fruit and vegetable consumption threshold [68]. Collectively, Canadians have been reporting a decreasing intake of fruits and vegetables, from 5.3 daily servings in 2004 to 4.5 daily servings in 2015, with a majority of people not meeting the guidelines in the 2007 Canada's food guide [69]. Fruit and vegetable intake among both immigrants and the collective North American population is therefore in need of improvement.

Additionally, studies in this review suggested that grains are an important source of energy for Chinese immigrants. Rice is especially regarded as a culturally significant food, as evidenced by its daily consumption and cultural associations [5,41,53]. Rice is regarded by Chinese immigrants as a symbolically comforting food with nuanced cultural and historical meanings [41]. However, the cultural importance of rice consumption has been identified to be a barrier in diabetes management, as higher consumption of white rice has been associated with increased risk of developing type 2 diabetes [41,70]. More research is necessary to quantify daily intake of such foods and their purported health effects and to better inform healthcare providers and the public's dietary decisions considering their cultural significance.

This review also found that consumption of milk and alternatives in Chinese immigrants was low, ranging from less than one serving per day to 1.6 servings/day. Chinese people in China consumed 11.8 g/day of milk, which also remains well below one serving [71]. By contrast, the consumption of milk and alternatives among Canadians is approximately 1.7 servings per day [72]. The Canadian food guide has moved away from strict serving guidelines for milk and alternatives, and instead encourages healthy proteins including low-fat dairy products [73]. These data would suggest increased efforts in choosing healthy dairy products and more attention to the source of protein rather than simply increasing intake of dairy. In addition, more research is needed on the adequacy of calcium consumption in Chinese immigrants, especially aging women prone to osteoporosis.

Findings from this review indicated that Chinese immigrants regularly consume meat and alternatives including soybeans, fish, and eggs, with higher consumption of the latter three compared to Caucasians. Although there is limited data on consumption of this food group among Chinese immigrants, the report of higher consumption of soybeans, fish, and eggs is in line with Canada's food guide's recommendations for protein sources, which also encourages plant-based proteins and lean meats. More data is needed to characterise meat and alternative consumption in Chinese immigrants, although the cultural preference for protein sources such as fish, eggs, and tofu presents itself as a facilitator of healthy dietary choices, as supported by Canada's food guide.

Carbohydrate intake for Chinese immigrants was reported by one study in this review to be 50% of daily calorie intake, which is within the American dietary guidelines of 45–65% of daily intake [67]. Fiber consumption was reported by several studies in this review, with values ranging from 12–14 g per day. This is less than the minimal recommended 25 g/day for women and 38 g/day for men intake, according to Canadian health guidelines [67,74,75]. This insufficient intake of fibre is consistent with previous research on Chinese adults in China, where daily fibre intake was 9.7 g/day, as well as Canadian statistics, with most Canadians only meeting half the recommended amount [74,76]. Specifically, for Chinese immigrants, fibre intake can be encouraged by promoting whole grain noodles and brown rice instead of white rice. Encouragement of fruits and vegetables, along with the aforementioned focus on ethno-cultural vegetables, can aid fibre intake.

Sufficient protein and fat consumption by Chinese immigrants against USDA dietary guidelines was demonstrated by studies in this review [62]. However, focus should be placed on plant-based proteins and minimising saturated fats. Encouragement of culturally relevant foods such as tofu, and certain vegetables, as well as encouragement of fat-reducing behaviours can help in choosing good sources of fats and proteins for Chinese immigrants.

Included studies from this review reported an increase in consumption of various food groups, adoption of Western food items, as well as maintenance of traditional diet components after immigration. These changes have both the potential for healthy and unhealthy impact on dietary behaviours. The increased intake of fruits and vegetables for example, would aid in the consumption of good sources of fibre, protein, and vitamins and minerals, especially considering the inadequate fibre consumption found in this population by this review. However, the increase in certain dietary behaviours such as an increased intake of refined carbohydrates, high-fat, and fast food is unsupported by the Canadian and USDA food guides, and thus is unhealthy [67,73]. Previous research on the healthy immigrant effect would suggest that immigrants, although healthier than native-born Canadians upon immigration, experience a decrease in health status that is partly attributable to dietary changes [10].

4.3. Influencing Factors

Findings from this review indicated that acculturation was associated with increased total energy intake, carbohydrate intake, and meat intake. These associations may contribute to concepts identified by previous research, where immigrants experience a decrease in health status that is partly attributable to dietary changes after immigration [10]. This would indicate a need to educate Chinese immigrants on distinguishing healthy versus unhealthy sources of carbohydrates and proteins, as their consumption is likely to increase with acculturation. There were conflicting findings with regards to the association between acculturation and overall dietary quality, fruit and vegetable consumption, and fat consumption. This may be in part due to inconsistencies in the measurement of acculturation. Studies used different indicators, such as number of American friends, media consumption, or English proficiency. The two-dimensional acculturation model where participants are non-exclusively assessed according to their maintenance of traditional culture and their adaptation to Western culture should be considered for future studies [77].

This review also indicated that individual factors influencing dietary behaviours of Chinese immigrants were associated with lack of time and nature of migration life experiences. The lack of time was identified as a barrier in other immigrant groups as well, such as Chinese immigrants in Spain as well as Arab Muslim immigrants to Canada [78,79]. The lack of time to prepare traditional meals leads to increased consumption of more easily prepared and accessible Western meals. Health promotion is necessary to facilitate healthy dietary decisions under time constraints, whether that involves choosing healthier Western alternatives, rather than refined carbohydrates and high-fat foods, or enabling easier access to healthy traditional foods. Interestingly, positive experiences also influenced healthier dietary behaviours, whereas stress was associated with a greater energy density and percentage of energy from fat. Stress related to immigration and acculturation may also be a contributing factor. This is consistent with previous research highlighting lack of social relationships, busier lifestyle, and higher stress levels as contributors to unhealthy dietary changes in immigrant women [80]. This highlights the importance of addressing stress and mental health among Chinese immigrants, as they are important determinants of dietary practices, which can in turn influence chronic health outcomes [80].

This review indicated that having young children in family influenced individual and familial dietary behaviors. Children's preferences have a particular influence on immigrant food choices, as it was identified as a factor in dietary choices of Chinese immigrants in Spain [79]. Children's preferences were also identified as an influencing factor in a study of Arab Muslim immigrant mothers in Canada [78]. Educational interventions on healthy dietary choices targeted at family members, specifically children, could encourage healthy eating behaviours in Chinese immigrants in North America.

Finally, this review indicated that cultural conceptualisations of health and eating as well as neighbourhood makeup were important community factors influencing Chinese immigrants' dietary behaviors. Cultural concepts of yin and yang, as well as hot and cold foods were important influences. Traditional Chinese Medicine dictates that certain foods are hot or cold by nature, and their balance is crucial in maintaining health [42]. Health promotion methods that account for both traditional Chinese conceptualisations of a healthy diet along with standard Western evidence-based models are necessary to promote healthy dietary behaviours among Chinese immigrants. Another interesting finding was that neighbourhoods with higher immigrant populations were associated with healthier consumption and lower high-fat and processed food consumption among Chinese immigrants. This finding is consistent among native-born Canadians, who showed healthier behaviours and better health outcomes in areas with higher immigrant density [81]. This indicates potential for immigrant food practices to be beneficial for the larger population as well [78,79].

4.4. Implications

There are several implications of our findings, especially on how they can be used to influence healthy eating, which should be aligned with current Canada Food Guide and USDA recommendations. Firstly, efforts can be made among Chinese immigrant individuals to increase their fruit and vegetable intake and fibre intake, resist unhealthy acculturative changes (i.e., increased consumption of high fat foods), and aid family members to do the same. Secondly, considering many Chinese immigrants seek nutritional information from family physicians [42], health care providers can enable healthy dietary changes using culturally sensitive methods. Diet plans and recommendations can include ethno-cultural vegetables and integrate traditional concepts where appropriate [82]. The sociocultural importance of certain staples and their contribution to well-being may also be recognised so that providers can work with individuals to find appropriate replacements or compromises. Healthcare providers can also take care to acknowledge and address the role of mental well-being in dietary decisions, with special consideration of the unique stressors immigrant groups may face. Thirdly, given their importance to this population, Chinese immigrant community organisations, especially those related to health, can play an important role in dietary education [83]. These can include cooking workshops or nutrition information sessions focused on encouraging fruits and vegetables, fibre, and avoiding saturated fats, among other healthy dietary behaviours. Finally, given that children are key influencers in the home diet, schools can create environments that support healthy eating and encourage children to bring home new health knowledge. Decisions regarding policies on immigration and neighbourhood planning can also consider the health benefits associated with areas of high immigrant density, improve access to ethnic foods, and promote sharing of cultural food practices. In conclusion, as has been proposed by previous studies, individuals, healthcare providers, and organisations can through their collective efforts, improve health outcomes for the population [84]. These findings and implications have the potential to be applicable to Chinese immigrants in other high-income Western countries, although the cultural, social, economic, and dietary differences between the countries should be considered prior to any generalisation of findings.

4.5. Limitations of This Review and Recommendations for Future Studies

There are some limitations in this review. Firstly, a variety of methods for measuring dietary intake are used across studies. Food groups and macronutrients were measured as servings, percentages, grams, and frequencies. The definition of a serving also varied, making it difficult to compare study findings. These discrepancies were further complicated by the fact that many studies grouped foods together (i.e., fruits and vegetables as one), while other studies did not. Furthermore, the Canada Food Guide and the USDA have different guidelines and food group categories, with the former providing no strict serving recommendations and no distinction between fruit and vegetable recommendations. Additionally, different food frequency questionnaires with different numbers of items and lengths of recall were used. Such methods rely heavily on participant memory, which is subjective and liable to error. Future studies should further develop the validity and reliability of food frequency questionnaires and other measurement tools to improve the rigor of dietary research. Secondly, there is a lack of studies undertaken on sodium, potassium, and calcium intake in Chinese immigrants. Given the importance of sodium and potassium levels in blood pressure regulation, and evidence of hypertension as the strongest risk factor for stroke in Chinese people, this aspect of dietary intake requires more investigation [85,86]. Dietary studies of calcium also need to be repeated in larger populations to discern whether the lower consumption of milk and alternatives among Chinese immigrants is a cause for concern. Finally, there is a lack of longitudinal cohort studies, case-control studies, and intervention studies on the nutrition of Chinese immigrants. The majority of data comes from cross-sectional studies relying heavily on self-reporting of dietary consumption over a few days. In the future, interventions studies, especially randomised controlled trials, should be conducted to investigate the role of promising dietary behaviours or specific dietary interventions on health outcomes in Chinese immigrants to North America.

5. Conclusions

This review characterises dietary behaviours and influencing factors of the Chinese immigrant population in Canada and the United States. This review found that fruit and vegetable, fibre, and dairy consumption among Chinese immigrants was generally insufficient. Although protein, fat, and carbohydrate intake was generally sufficient, efforts can be taken to ensure healthy sources are selected. Dietary acculturation was also observed in the Chinese immigrant population in this review, and although not an inherently negative change, efforts can be undertaken to ensure that healthy Western foods are adopted. Dietary behaviours in Chinese immigrants are influenced by everything from traditional health beliefs, time, and accessibility, to family members and neighbourhoods. It is important for healthcare providers and nutritionists to remain culturally sensitive when providing dietary recommendations. This can be achieved through encouragement of healthy ethnocultural foods and incorporating traditional health beliefs into Western evidence-based principles.

Author Contributions: Conceptualization, P.Z., Y.L., Y.Y., C.Z., H.Z. and Y.W.; methodology, P.Z. and Y.W.; formal analysis, P.Z. and D.B.; data curation, P.Z. and D.B.; writing—original draft preparation, P.Z. and D.B.; writing—review and editing, all authors; supervision, P.Z.; funding acquisition, P.Z. All authors have read and agreed to the published version of the manuscript.

Funding: This research was funded by the Social Sciences and Humanities Research Council Institutional Grant (Ref # 102830).

Informed Consent Statement: Not Applicable.

Conflicts of Interest: The authors declared no conflict of interest.

References

1. Government of Canada. *Immigration and Ethnocultural Diversity Statistics*; Statistics Canada: Ottawa, ON, Canada, 2019. Available online: https://www.statcan.gc.ca/en/subjects-start/immigration_and_ethnocultural_diversity (accessed on 20 May 2022).
2. Estrada, C.E.; Batalova, J. Chinese Immigrants in the United States. 2020. Available online: migrationpolicy.org (accessed on 20 May 2022).
3. Suryadinata, L. *Migration, Indigenization, and Interaction: Chinese Overseas and Globalization*; World Scientific: Singapore, 2011; 335p.
4. Pearce, R.R.; Lin, Z. Chinese American post-secondary achievement and attainment: A cultural and structural analysis. *Educ. Rev.* **2007**, *59*, 19–36. [CrossRef]
5. Washington, G.; Wang-Letzkus, M.F. Self-care practices, health beliefs, and attitudes of older diabetic Chinese Americans. *J. Health Hum. Serv. Adm.* **2009**, *32*, 305–323. [PubMed]
6. Nestler, G. Traditional Chinese medicine. *Med. Clin.* **2002**, *86*, 63–73. [CrossRef]
7. Higginbottom, G.M.A.; Vallianatos, H.; Shankar, J.; Safipour, J.; Davey, C. "Immigrant women's food choices in pregnancy: Perspectives from women of Chinese origin in Canada": Corrigendum. *Ethn. Health* **2018**, *23*, 914. [CrossRef] [PubMed]
8. Beasley, J.M.; Yi, S.S.; Ahn, J.; Kwon, S.C.; Wylie-Rosett, J. Dietary Patterns in Chinese Americans are Associated with Cardiovascular Disease Risk Factors, the Chinese American Cardiovascular Health Assessment (CHA CHA). *J. Immigr. Minor. Health* **2019**, *21*, 1061–1069. [CrossRef]
9. Roberts, C.K.; Barnard, R.J. Effects of exercise and diet on chronic disease. *J. Appl. Physiol.* **2005**, *98*, 3–30. [CrossRef]
10. Sanou, D.; O'Reilly, E.; Ngnie-Teta, I.; Batal, M.; Mondain, N.; Andrew, C.; Newbold, B.K.; Bourgeault, I. Acculturation and Nutritional Health of Immigrants in Canada: A Scoping Review. *J. Immigr. Minor. Health* **2014**, *16*, 24–34. [CrossRef]
11. LeCroy, M.N.; Stevens, J. Dietary intake and habits of South Asian immigrants living in Western countries. *Nutr. Rev.* **2017**, *75*, 391–404. [CrossRef]

12. Elshahat, S.; Moffat, T. Dietary practices among Arabic-speaking immigrants and refugees in Western societies: A scoping review. *Appetite* **2020**, *154*, 104753. [CrossRef]

13. Deng, F.; Zhang, A.; Chan, C.B. Acculturation, Dietary Acceptability, and Diabetes Management among Chinese in North America. *Front. Endocrinol.* **2013**, *4*, 108. [CrossRef]

14. Li-Geng, T.; Kilham, J.; McLeod, K.M. Cultural Influences on Dietary Self-Management of Type 2 Diabetes in East Asian Americans: A Mixed-Methods Systematic Review. *Health Equity* **2020**, *4*, 31–42. [CrossRef] [PubMed]

15. Krieger, N. Theories for social epidemiology in the 21st century: An ecosocial perspective. *Int. J. Epidemiol.* **2001**, *30*, 668–677. [CrossRef] [PubMed]

16. Zou, P. Facilitators and Barriers to Healthy Eating in Aged Chinese Canadians with Hypertension: A Qualitative Exploration. *Nutrients* **2019**, *11*, 111. [CrossRef] [PubMed]

17. Tricco, A.C.; Lillie, E.; Zarin, W.; O'Brien, K.K.; Colquhoun, H.; Levac, D.; Moher, D.; Peters, M.D.; Horsley, T.; Weeks, L.; et al. PRISMA Extension for Scoping Reviews (PRISMA-ScR): Checklist and Explanation. *Ann. Intern. Med.* **2018**, *169*, 467–473. [CrossRef] [PubMed]

18. Critical Appraisal Skills Programme. CASP Checklists. 2017. Available online: http://www.casp-uk.net/casp-tools-checklists (accessed on 5 May 2022).

19. A Rodriguez, L.; Jin, Y.; A Talegawkar, S.; Otto, M.C.D.O.; Kandula, N.R.; Herrington, D.M.; Kanaya, A.M. Differences in Diet Quality among Multiple US Racial/Ethnic Groups from the Mediators of Atherosclerosis in South Asians Living in America (MASALA) Study and the Multi-Ethnic Study of Atherosclerosis (MESA). *J. Nutr.* **2020**, *150*, 1509–1515. [CrossRef]

20. Chai, S.C.; Jiang, H.; Papas, M.A.; Fang, C.-S.; Setiloane, K.T. Acculturation, diet, and psychological health among Asian students. *J. Am. Coll. Health* **2019**, *67*, 433–440. [CrossRef]

21. Kirshner, L.; Yi, S.; Wylie-Rosett, J.; Matthan, N.R.; Beasley, J. Acculturation and Diet Among Chinese American Immigrants in New York City. *Curr. Dev. Nutr.* **2019**, *4*, nzz124. [CrossRef]

22. An, Z. Diet-specific social support, dietary acculturation, and self-efficacy among Chinese living in the United States. *J. Int. Intercult. Commun.* **2018**, *11*, 136–153. [CrossRef]

23. Liu, L.W. From shopping to eating: Food safety practices among Chinese immigrants in Canada. *Int. J. Anthr. Ethnol.* **2017**, *1*, 4. [CrossRef]

24. Lu, C.; McGinn, M.K.; Xu, X.; Sylvestre, J. Living in Two Cultures: Chinese Canadians' Perspectives on Health. *J. Immigr. Minor. Health* **2017**, *19*, 423–429. [CrossRef]

25. Zou, P.; Dennis, C.-L.; Cindy-Lee, D.; Parry, M. Hypertension Prevalence, Health Service Utilization, and Participant Satisfaction: Findings From a Pilot Randomized Controlled Trial in Aged Chinese Canadians. *Inq. J. Health Care Organ. Provis. Financ.* **2017**, *54*, 46958017724942. [CrossRef]

26. Yi, S.S.; Thorpe, L.E.; Zanowiak, J.M.; Trinh-Shevrin, C.; Islam, N.S. Clinical Characteristics and Lifestyle Behaviors in a Population-Based Sample of Chinese and South Asian Immigrants With Hypertension. *Am. J. Hypertens.* **2016**, *29*, 941–947. [CrossRef] [PubMed]

27. Tseng, M.; Wright, D.J.; Fang, C.Y. Acculturation and Dietary Change Among Chinese Immigrant Women in the United States. *J. Immigr. Minor. Health* **2015**, *17*, 400–407. [CrossRef] [PubMed]

28. Corlin, L.; Woodin, M.; Thanikachalam, M.; Lowe, L.; Brugge, D. Evidence for the healthy immigrant effect in older Chinese immigrants: A cross-sectional study. *BMC Public Health* **2014**, *14*, 603. [CrossRef] [PubMed]

29. Wyatt, L.C.; Trinh-Shevrin, C.; Islam, N.S.; Kwon, S. Health-Related Quality of Life and Health Behaviors in a Population-Based Sample of Older, Foreign-Born, Chinese American Adults Living in New York City. *Health Educ. Behav.* **2014**, *41* (Suppl. 1), 98S–107S. [CrossRef] [PubMed]

30. Adekunle, B.; Filson, G.; Sethuratnam, S. Immigration and Chinese food preferences in the Greater Toronto Area. *Int. J. Consum. Stud.* **2013**, *37*, 658–665. [CrossRef]

31. Wong, S.S.; Dixon, L.B.; Gilbride, J.A.; Kwan, T.W.; Stein, R.A. Measures of Acculturation are Associated with Cardiovascular Disease Risk Factors, Dietary Intakes, and Physical Activity in Older Chinese Americans in New York City. *J. Immigr. Minor. Health* **2013**, *15*, 560–568. [CrossRef]

32. Tam, C.Y.; Hislop, G.; Hanley, A.J.; Minkin, S.; Boyd, N.F.; Martin, L.J. Food, Beverage, and Macronutrient Intakes in Post-menopausal Caucasian and Chinese-Canadian Women. *Nutr. Cancer* **2011**, *63*, 687–698. [CrossRef]

33. Tseng, M.; Fang, C.Y. Socio-economic position and lower dietary moderation among Chinese immigrant women in the USA. *Public Health Nutr.* **2012**, *15*, 415–423. [CrossRef]

34. Alonge, O.K.; Narendran, S.; Hobdell, M.H.; Bahl, S. Sugar consumption and preference among Mexican, Chinese, and Nigerian immigrants to Texas. *Speéc. Care Dent.* **2011**, *31*, 150–155. [CrossRef]

35. Lv, N.; Brown, J.L. Impact of a Nutrition Education Program to Increase Intake of Calcium-Rich Foods by Chinese-American Women. *J. Am. Diet. Assoc.* **2011**, *111*, 143–149. [CrossRef] [PubMed]

36. Rosenmöller, D.L.; Gasevic, D.; Seidell, J.; A Lear, S. Determinants of changes in dietary patterns among Chinese immigrants: A cross-sectional analysis. *Int. J. Behav. Nutr. Phys. Act.* **2011**, *8*, 42. [CrossRef] [PubMed]

37. Tam, C.Y.; Martin, L.J.; Hislop, G.; Hanley, A.J.; Minkin, S.; Boyd, N.F. Risk factors for breast cancer in postmenopausal Caucasian and Chinese-Canadian women. *Breast Cancer Res.* **2010**, *12*, R2. [CrossRef] [PubMed]

38. Tseng, M.; Fang, C.Y. Stress Is Associated with Unfavorable Patterns of Dietary Intake Among Female Chinese Immigrants. *Ann. Behav. Med.* **2011**, *41*, 324–332. [CrossRef] [PubMed]

39. Liu, A.; Berhane, Z.; Tseng, M. Improved Dietary Variety and Adequacy but Lower Dietary Moderation with Acculturation in Chinese Women in the United States. *J. Am. Diet. Assoc.* **2010**, *110*, 457. [CrossRef]

40. Bell, K.; Lee, J.; Ristovski-Slijepcevic, S. Perceptions of food and eating among Chinese patients with cancer: Findings of an ethnographic study. *Cancer Nurs.* **2009**, *32*, 118–126. [CrossRef]

41. Chesla, C.A.; Chun, K.M.; Kwan, C.M. Cultural and Family Challenges to Managing Type 2 Diabetes in Immigrant Chinese Americans. *Diabetes Care* **2009**, *32*, 1812–1816. [CrossRef]

42. Kwok, S.; Mann, L.; Wong, K.; Blum, I. Dietary habits and health beliefs of Chinese Canadians. *Can. J. Diet. Pr. Res.* **2009**, *70*, 73–80. [CrossRef]

43. Osypuk, T.L.; Roux, A.V.D.; Hadley, C.; Kandula, N.R. Are immigrant enclaves healthy places to live? The Multi-ethnic Study of Atherosclerosis. *Soc. Sci. Med.* **2009**, *69*, 110–120. [CrossRef]

44. Hislop, T.G.; Tu, S.-P.; Teh, C.; Li, L.; Low, A.; Taylor, V.M.; Yasui, Y. Knowledge and Behaviour Regarding Heart Disease Prevention in Chinese Canadian Immigrants. *Can. J. Public Health* **2008**, *99*, 232–235. [CrossRef]

45. Kandula, N.R.; Diez-Roux, A.V.; Chan, C.; Daviglus, M.L.; Jackson, S.A.; Ni, H.; Schreiner, P.J. Association of Acculturation Levels and Prevalence of Diabetes in the Multi-Ethnic Study of Atherosclerosis (MESA). *Diabetes Care* **2008**, *31*, 1621–1628. [CrossRef] [PubMed]

46. Lu, C.; Sylvestre, J.; Melnychuk, N.; Li, J. East Meets West: Chinese-Canadians' Perspectives on Health and Fitness. *Can. J. Public Health* **2008**, *99*, 22–25. [CrossRef] [PubMed]

47. Taylor, V.M.; Yasui, Y.; Tu, S.P.; Neuhouser, M.L.; Li, L.; Woodall, E.; Acorda, E.; Cripe, S.M.; Hislop, T.G. Heart disease prevention among Chinese immigrants. *J. Community Health: Publ. Health Promot. Dis. Prev.* **2007**, *32*, 299–310. [CrossRef] [PubMed]

48. Babbar, R.K.; Handa, A.B.; Lo, C.-M.; Guttmacher, S.J.; Shindledecker, R.; Chung, W.; Fong, C.; Ho-Asjoe, H.; Chan-Ting, R.; Dixon, L.B. Bone Health of Immigrant Chinese Women Living in New York City. *J. Community Health* **2006**, *31*, 7–23. [CrossRef]

49. Fang, J.; Foo, S.H.; Fung, C.; Wylie-Rosett, J.; Alderman, M.H. Stroke Risk among Chinese Immigrants in New York City. *J. Immigr. Minor. Health* **2006**, *8*, 387–393. [CrossRef]

50. Walker, M.D.; Babbar, R.; Opotowsky, A.; McMahon, D.J.; Liu, G.; Bilezikian, J.P. Determinants of bone mineral density in Chinese-American women. *Osteoporos. Int.* **2007**, *18*, 471–478. [CrossRef]

51. Liang, W.; Yuan, E.; Mandelblatt, J.S.; Pasick, R.J. How do older Chinese women view health and cancer screening? results from focus groups and implications for interventions. *Ethn. Health* **2004**, *9*, 283–304. [CrossRef]

52. Lv, N.; Cason, K.L. Dietary pattern change and acculturation of Chinese Americans in Pennsylvania. *J. Am. Diet. Assoc.* **2004**, *104*, 771–778. [CrossRef]

53. Lv, N.; Cason, K.L. Current Dietary Pattern and Acculturation of Chinese Americans in Pennsylvania. *Top. Clin. Nutr.* **2003**, *18*, 291–300. [CrossRef]

54. Kelemen, L.E.; Anand, S.S.; Vuksan, V.; Yi, Q.; Teo, K.K.; Devanesen, S.; Yusuf, S. Development and evaluation of cultural food frequency questionnaires for South Asians, Chinese, and Europeans in North America. *J. Am. Diet. Assoc.* **2003**, *103*, 1178–1184. [CrossRef]

55. Satia-Abouta, J.; Patterson, R.E.; Kristal, A.; Teh, C.; Tu, S.-P. Psychosocial Predictors of Diet and Acculturation in Chinese American and Chinese Canadian Women. *Ethn. Health* **2002**, *7*, 21–39. [CrossRef] [PubMed]

56. Wu, A.H.; Wan, P.; Hankin, J.; Tseng, C.-C.; Yu, M.C.; Pike, M.C. Adolescent and adult soy intake and risk of breast cancer in Asian-Americans. *Carcinogenesis* **2002**, *23*, 1491–1496. [CrossRef] [PubMed]

57. Liou, D.; Contento, I.R. Usefulness of Psychosocial Theory Variables in Explaining Fat-Related Dietary Behavior in Chinese Americans: Association with Degree of Acculturation. *J. Nutr. Educ.* **2001**, *33*, 322–331. [CrossRef]

58. A Satia, J.; E Patterson, R.; Kristal, A.R.; Hislop, T.G.; Pineda, M. A household food inventory for North American Chinese. *Public Health Nutr.* **2001**, *4*, 241–247. [CrossRef]

59. A Satia, J.; E Patterson, R.; Kristal, A.R.; Hislop, T.; Yasui, Y.; Taylor, V.M. Development of scales to measure dietary acculturation among Chinese-Americans and Chinese-Canadians. *J. Am. Diet. Assoc.* **2001**, *101*, 548–553. [CrossRef]

60. Satia, J.A.; Patterson, R.E.; Taylor, V.M.; Cheney, C.L.; Shiu-Thornton, S.; Chitnarong, K.; Kristal, A.R. Use of qualitative methods to study diet, acculturation, and health in Chinese-American women. American Dietetic Association. *J. Am. Diet. Assoc.* **2000**, *100*, 934–940. [CrossRef]

61. Whittemore, A.S.; Kolonel, L.N.; Wu, A.H.; John, E.M.; Gallagher, R.P.; Howe, G.R.; Burch, J.D.; Hankin, J.; Dreon, D.M.; West, D.W.; et al. Prostate Cancer in Relation to Diet, Physical Activity, and Body Size in Blacks, Whites, and Asians in the United States and Canada. *JNCI J. Natl. Cancer Inst.* **1995**, *87*, 652–661. [CrossRef]

62. Choi, E.S.K. The Prevalence of Cardiovascular Risk Factors Among Elderly Chinese Americans. *Arch. Intern. Med.* **1990**, *150*, 413–418. [CrossRef]

63. Whittemore, A.S.; Wu-Williams, A.H.; Lee, M.; Zheng, S.; Gallagher, R.P.; A Jiao, D.; Zhou, L.; Wang, X.H.; Chen, K.; Jung, D. Diet, physical activity, and colorectal cancer among Chinese in North America and China. *JNCI J. Natl. Cancer Inst.* **1990**, *82*, 915–926. [CrossRef]

64. Newman, J.M. *Chinese Immigrant Food Habits: A Study of the Nature and Direction of Change*; New York University: Ann Arbor, MI, USA, 1980; p. 149.

65. Wang, K. *Acculturation, Sociodemographic and Environmental Determinants of Dietary Intake among Asian Immigrants in the United States*; Graduate School of Social Work, Boston University: Boston, MA, USA, 2017.

66. Milani, G.P.; Silano, M.; Pietrobelli, A.; Agostoni, C. Junk food concept: Seconds out. *Int. J. Obes.* **2017**, *41*, 669. [CrossRef]

67. U.S. Department of Agriculture; U.S. Department of Health and Human Services. *Dietary Guidelines for Americans, 2020–2025*, 9th ed.; US Government Printing Office: Washington, DC, USA, 2020; Volume 164.

68. Dai, C.-L.; Sharma, M.; Haider, T.; Sunchu, H. Fruit and Vegetable Consumption Behavior Among Asian Americans: A Thematic Analysis. *J. Prim. Care Community Health* **2021**, *12*, 2150132720984776. [CrossRef] [PubMed]

69. Polsky, J.Y.; Garriguet, D. Change in vegetable and fruit consumption in Canada between 2004 and 2015. *Health Rep.* **2020**, *31*, 3–12. [PubMed]

70. A Hu, E.; Pan, A.; Malik, V.; Sun, Q. White rice consumption and risk of type 2 diabetes: Meta-analysis and systematic review. *BMJ* **2012**, *344*, e1454. [CrossRef] [PubMed]

71. He, Y.; Yang, X.; Xia, J.; Zhao, L.; Yang, Y. Consumption of meat and dairy products in China: A review. *Proc. Nutr. Soc.* **2016**, *75*, 385–391. [CrossRef]

72. Auclair, O.; Han, Y.; Burgos, S.A. Consumption of Milk and Alternatives and Their Contribution to Nutrient Intakes among Canadian Adults: Evidence from the 2015 Canadian Community Health Survey—Nutrition. *Nutrients* **2019**, *11*, 1948. [CrossRef]

73. Government of Canada. *What Are Canada's Dietary Guidelines? Canada Food Guide*; Health Canada: Ottawa, ON, Canada, 2020. Available online: https://food-guide.canada.ca/en/ (accessed on 20 May 2022).

74. Government of Canada. *Nutrients in Food*; Health Canada: Ottawa, ON, Canada, 2015. Available online: https://www.canada.ca/en/health-canada/services/nutrients.html (accessed on 20 May 2022).

75. Government of Canada. *Fibre*; Health Canada: Ottawa, ON, Canada, 2012. Available online: https://www.canada.ca/en/health-canada/services/nutrients/fibre.html (accessed on 20 May 2022).

76. Yu, D.; Zhao, L.; Zhao, W. Status and trends in consumption of grains and dietary fiber among Chinese adults (1982–2015). *Nutr. Rev.* **2020**, *78* (Suppl. 1), 43–53. [CrossRef]

77. Worthy, L.D.; Lavigne, T.; Romero, F. Berry's Model of Acculturation. In *Culture and Psychology*; Glendale Community College: Phoenix, AZ, USA, 2020.

78. Aljaroudi, R.; Horton, S.; Hanning, R.M. Acculturation and Dietary Acculturation among Arab Muslim Immigrants in Canada. *Can. J. Diet. Pract. Res.* **2019**, *80*, 172–178. [CrossRef]

79. Benazizi, I.; Ferrer-Serret, L.; Martínez-Martínez, J.M.; Ronda-Pérez, E.; Barbarà, C.I. Factors that influence the diet and eating habits of Chinese immigrant population in Catalonia (Spain). *Gac. Sanit.* **2021**, *35*, 12–20. [CrossRef]

80. Popovic-Lipovac, A.; Strasser, B. A Review on Changes in Food Habits Among Immigrant Women and Implications for Health. *J. Immigr. Minor. Health* **2015**, *17*, 582–590. [CrossRef]

81. Shi, L.; Zhang, D.; Rajbhandari-Thapa, J.; Katapodis, N.; Su, D.; Li, Y. Neighborhood immigrant density and population health among native-born Americans. *Prev. Med.* **2019**, *127*, 105792. [CrossRef]

82. Lee, M.M.; Shen, J.M. Dietary patterns using Traditional Chinese Medicine principles in epidemiological studies. *Asia Pac. J. Clin. Nutr.* **2008**, *17* (Suppl. 1), 79–81. [PubMed]

83. Zhou, M.; Lee, R. Transnationalism and Community Building: Chinese Immigrant Organizations in the United States. *ANNALS Am. Acad. Political Soc. Sci.* **2013**, *647*, 22–49. [CrossRef]

84. Valente, M.; Cortesi, P.A.; Lassandro, G.; Mathew, P.; Pocoski, J.; Molinari, A.C.; Mantovani, L.G.; Giordano, P. Health economic models in hemophilia A and utility assumptions from a clinician's perspective. *Pediatric Blood Cancer* **2015**, *62*, 1826–1831. [CrossRef] [PubMed]

85. Mozaffarian, D.; Fahimi, S.; Singh, G.M.; Micha, R.; Khatibzadeh, S.; Engell, R.E.; Lim, S.; Danaei, G.; Ezzati, M.; Powles, J.; et al. Global Sodium Consumption and Death from Cardiovascular Causes. *N. Engl. J. Med.* **2014**, *371*, 624–634. [CrossRef] [PubMed]

86. Wang, J.; Wen, X.; Li, W.; Li, X.; Wang, Y.; Lu, W. Risk Factors for Stroke in the Chinese Population: A Systematic Review and Meta-Analysis. *J. Stroke Cerebrovasc. Dis.* **2017**, *26*, 509–517. [CrossRef]

nutrients

Article

"Food Is Our Love Language": Using Talanoa to Conceptualize Food Security for the Māori and Pasifika Diaspora in South-East Queensland, Australia

Heena Akbar [1,2], Charles J. T. Radclyffe [1,2,3], Daphne Santos [4], Maureen Mopio-Jane [5] and Danielle Gallegos [1,2,*]

1   Woolworths Centre for Childhood Nutrition Research, Queensland University of Technology, South Brisbane 4101, Australia; h.akbar@qut.edu.au (H.A.); c_radclyffe93@hotmail.com (C.J.T.R.)
2   School of Exercise and Nutrition Sciences, Queensland University of Technology, Kelvin Grove 4059, Australia
3   Pasifika Young Peoples Well-Being Network (PYPWN), School of Public Health and Social Work, Queensland University of Technology, Kelvin Grove 4059, Australia
4   Good Start Program, Child and Youth Community Health Services, Children's Health Queensland, South Brisbane 4101, Australia; daphne.santos@health.qld.gov.au
5   Elder of the Papua New Guinea Community, Queensland 4000, Australia; mmopiojane@me.com
*   Correspondence: danielle.gallegos@qut.edu.au

Abstract: Queensland is home to the largest diaspora of Māori and Pasifika peoples in Australia. They form an understudied population concerning experiences and challenges of food insecurity. This community co-designed research aims to explore the conceptualization of household food security by Māori and Pasifika peoples living in south-east Queensland. Participatory action research and talanoa were used to collect and analyse forty interviews with leaders representing 22 Māori and Pasifika cultural identities in south-east Queensland. Eight key themes emerged that conceptualise food security as an integral part of the culture and holistic health. These themes included: spirituality, identity, hospitality and reciprocity, stigma and shame, expectations and obligations, physical and mental health and barriers and solutions. Addressing food insecurity for collectivist cultures such as Māori and Pasifika peoples requires embracing food sovereignty approaches for improved food security through the co-design of practical solutions that impact social determinants and strengthen existing networks to produce and distribute affordable and nutritious food.

Keywords: food security; culture; food sovereignty; monitoring; Pacific Islands; Māori; Melanesian; Micronesian; Polynesian

Citation: Akbar, H.; Radclyffe, C.J.T.; Santos, D.; Mopio-Jane, M.; Gallegos, D. "Food Is Our Love Language": Using Talanoa to Conceptualize Food Security for the Māori and Pasifika Diaspora in South-East Queensland, Australia. *Nutrients* **2022**, *14*, 2020. https://doi.org/10.3390/nu14102020

Academic Editors: Colin Bell and Penny Love

Received: 12 April 2022
Accepted: 9 May 2022
Published: 11 May 2022

Publisher's Note: MDPI stays neutral with regard to jurisdictional claims in published maps and institutional affiliations.

1. Introduction

Food security exists "when all people at all times have physical, social, and economic access to sufficient, safe and nutritious food to meet their dietary needs and food preferences for an active and healthy life" [1]. Being able to feed yourself and your family with food that is of sufficient quantity and quality is a fundamental human right [2]. In high-income countries, such as Australia, food is often available in sufficient quantities, but access to a diet of sufficient quality to sustain health can be compromised [3]. A recent review has also highlighted that culture, including but not limited to religiosity, gender, perceptions of stigma and food sharing norms, potentially impacts on every element of food security [4].

Food insecurity at the household level has serious implications for the physical and psycho-social health and well-being of adults and children. Food insecure adults and children have increased healthcare use and presentations to emergency departments [5–8]. Adults experiencing household food insecurity (HFI) have a greater risk of chronic conditions (diabetes, heart disease) and mortality [9–12] as well as poor mental health [13–15]. In children, HFI has been associated with compromised mental health [16], as well as developmental, cognitive and behavioural (internalising and externalising) anomalies [17]. In parts of Australia, from one in four to one in two households experience food insecurity on a regular basis [7,18–22]. The primary determinant of HFI is low income combined with the high cost of living, including unaffordable housing, utilities, healthcare, education, and food [3,23–25]. Other contributing factors include the number of adults and children in a household, mental and physical health of adults and children, psychological stress, adverse childhood experiences, domestic and family violence and experiences of racism [26–30].

Māori and Pasifika peoples currently make up an estimated 102,320 of the Queensland population, which is the largest Māori and Pacific Islander population in Australia [31]. Due to a complex interplay between cultural and social determinants as well as migration conditions, Māori and Pasifika peoples living in Queensland are at high risk of food insecurity. The next section provides background into the unique position of the Māori and Pasifika diaspora living in Australia.

1.1. Māori and Pasifika in Australia

The term 'Pasifika' is used in this study to describe the intergenerational diaspora of migrants who belong to and identify with Oceania which encompasses all Pacific Island nations that constitute Micronesia, Melanesia and Polynesia [32]. The term acknowledges and celebrates the layered and rich diversity of this oceanic continent; however, it recognises that there are sufficient commonalities across beliefs, values, and practices [33]. 'Pasifika' has been used in other Australian health studies and accepted by some communities to represent the Māori (tangata whenua—"people of the land") of Aotearoa (New Zealand) [32,33]. However, the communities in this research requested the use of 'Māori and Pasifika' separately in recognition that many Pasifika have New Zealand citizenship, but Māori are the First Nations people of Aotearoa.

Māori and Pasifika communities in Queensland are well connected and work collectively through a rich network of cultural, sporting, religious, educational, and support organisations [33,34]. The extended network is the location for the cultivation and enactment of identity, spirituality, social principles, and responsibilities [35,36]. Despite these considerable strengths, Māori and Pasifika peoples are impacted by a combination of pre-existing social and cultural determinants and the precarity engendered by the immigration stance of the Australian government. This means that Māori and Pasifika peoples are more likely to work in unskilled or semi-skilled positions and be at risk of irregular, casualised employment due to racism, language ability and low education attainment [37,38]. There is a lack of recent data, but in Australia and Queensland specifically, Māori and Pasifika peoples are more likely to be hospitalised and die prematurely from chronic conditions such as diabetes [39,40]. The high risk and prevalence of chronic conditions are exacerbated by poor diet quality [41,42]. Diet quality can be difficult to maintain due to cultural food insecurity, where the food preferences of the community cannot be realised due to those foods being unavailable or too expensive [43].

One of the most significant developments in Māori and Pasifika migration between New Zealand, Australia and Pacific Island nations was the establishment of the Trans-Tasman Travel Arrangement (TTTA) in 1973. The TTTA is firmly rooted in Australian attempts to reduce perceived "back door" migration to Australia [36,44]. Changes to the TTTA have resulted in visa holders after 2001 being ineligible for social security with only limited access to the National Disability Scheme, public and emergency housing, transport concessions and other assistance [45]. The system creates a lack of equity, high rates of unemployment, under-employment and precarious employment contributing to low incomes, all of which contribute to food insecurity.

Food insecurity is therefore potentially a significant issue for this population due to the financial hardships generated by the inability to access social protection, public housing and other social services as well as increased vulnerability to the escalating cost of living and shocks such as adverse weather events. This is further exacerbated by the sending of remittances to Island communities [46]. All of these factors contribute to family breakdown, over-representation in the justice system [36], and overall poorer health [40]. It is therefore essential to explore how food insecurity is conceptualized in this community in order to co-design strategies in the future.

1.2. Aims

The aim of this research was to explore how Māori and Pasifika communities living in south-east Queensland, conceptualise and perceive food security. The results will enable an in-depth understanding of the significance of food beyond physical health and contribute to co-designing solutions to ensure food security for Māori and Pasifika communities.

2. Materials and Methods

This research is reported using the Standards for Reporting Qualitative Research (SRQR) guidelines (see Supplementary Materials S1).

2.1. Design

This qualitative research was underpinned by the principles of a rights-based approach to food, focusing on the alteration of conditions and environments to enable people to take an active role in procuring food [47]. Participatory action research (PAR) utilising co-design and talanoa methods was employed as the theoretical and cultural framework in the collection and analysis of data. 'Talanoa' is a Pasifika method/methodology describing a process of enabling critical discussions and knowledge co-construction. 'Tala' means to inform or relate and 'noa' means to share as an exchange without a rigid framework [48].

The project was led by a Steering Committee (SC) including Māori and Pasifika members living in south-east Queensland, a Pasifika university researcher (HA) and an experienced university researcher (DG). The SC was guided by a co-developed terms of reference that outlined expectations and parameters of engagement. The SC decided the data to be collected and analysed by community researchers (CR) drawn from the Pasifika communities. Six Elders and two young researchers (<30 years) volunteered from the SC and an additional Elder researcher and five young researchers were recruited from the communities and the Pacific Young Peoples Wellbeing Network (PYPWN). Each young researcher was paired with an Elder researcher to optimise the knowledge exchange and to promote intergenerational conversations regarding culture and food. All CRs received training in culturally safe data collection, ethics (including confidentiality), data management and analysis within the context of talanoa and were remunerated as research assistants.

2.2. Participants and Recruitment

The SC brainstormed members of the Māori and Pasifika communities who held positions of high esteem or were known leaders, had advocated on food insecurity or were developing interventions related to food security (e.g., food pantries, gardens). Key recruitment considerations included capturing a diverse representation from all Pacific Island nations (irrespective of the size of the community), from multiple generations and those born outside and in Australia. Community members were approached initially by an SC member, either by telephone or email, to gauge interest to participate in the study. Once they had agreed, participant information was sent and follow up was made by the HA. This allowed community members to agree or decline to participate and this process reduced the risk of perceived coercion. HA emailed the participant information, consent, and the interview process with each participant and answered any questions regarding the research. Finally, participants were able to declare a preference for an interviewer allowing the agency to request or decline a particular CR. This enabled the participant to feel comfortable if sensitive material was to be shared. Interview day, time, location and mode were organised by the CRs and their assigned participant.

2.3. Data Collection

The SC iteratively developed the talanoa questions which were further refined in consultation with the CR. The questions and talanoa were then piloted with members of the SC to ensure they were clearly understood, culturally appropriate and elicited the information expected based on the research questions. The interview guide is located in Supplementary Materials S2.

Prior to each talanoa, the CR went through the participant information again, answering any additional questions and gained written or verbal consent. Cultural sensitivity and respect through talanoa formed the underpinning guiding value in this research process [48]. Each participant received an AUD 30 dollar grocery voucher as koha or a gift for participation.

Data collection occurred over a three-month period from September to November 2020 and interviews were conducted face-to-face or via the online video communication platform Zoom (©2021 Zoom Video Communications, Inc., San Jose, CA, USA), based on participant preference. Talanoa sessions took between 90 to 120 min. All interviews were audio-recorded and transcribed using Otter ai 2.0 Software (Otter.ai. Software Development, Mountain View, CA, USA). Each CR team uploaded the recording onto a secured shared drive, accessible only to the university researchers. Participants were given the opportunity to read a copy of their transcript and make changes. Each interview transcript was reviewed by the lead researchers for verbatim accuracy against the recording, making sure that any words in the language were captured correctly. Words that were incorrectly transcribed were reviewed with the CR and SC and edited accordingly.

2.4. Data Co-Analysis

Figure 1 summarises the data analysis process.

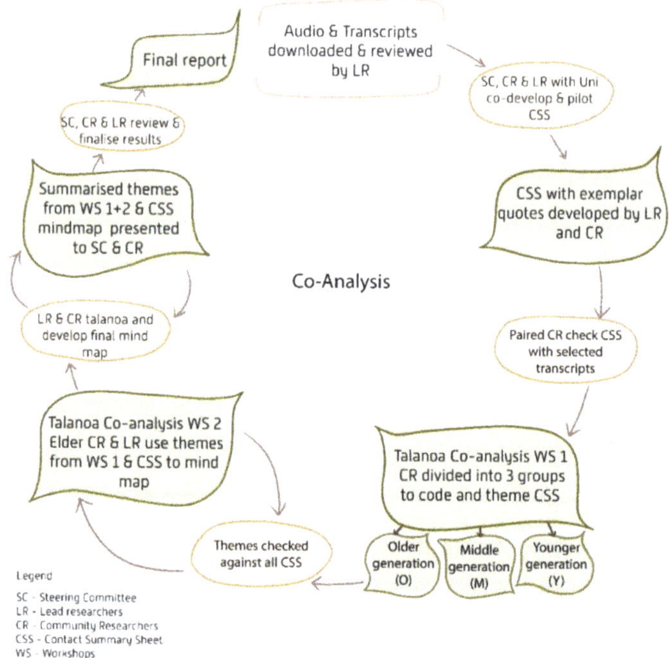

Figure 1. Data analysis process.

Due to the volume and depth of data collected and the logistics of involving 16 researchers in the data analysis, the SC and university researchers (HA and DG) co-developed a contact summary sheet (CSS) [49]. A CSS is a single sheet containing a series of focusing or summarizing questions about a particular interview (see Supplementary Materials S3).

The CSS was iteratively developed with the SC and CR and included verbatim quotes. Community and university researchers independently listened to and read one interview audio recording and transcript to complete the CSS. Any differences were discussed to develop a shared understanding of the material. University researchers completed the remaining transcripts. The CSS was used in conjunction with the interview transcripts for coding and thematic analysis. To aid in interpreting the large datasets, each transcript was uploaded to an online wordle tool (www.edwordle.net accessed on 29 April 2021). The wordle was set to 250 words, to group similar words and exclude stop words (Figure 2).

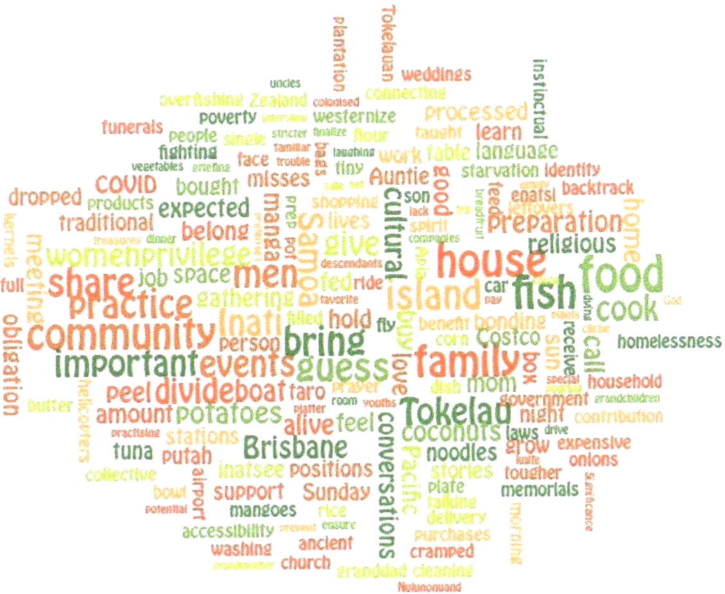

Figure 2. Wordle example from one of the interviews.

Thematic Analysis

Line-by-line coding and inductive thematic analysis were undertaken by the university researchers using the steps outlined by Braun and Clarke to develop a preliminary codebook [50]. The preliminary codebook, transcripts, CSS and wordles were used in a talanoa workshop with CR. This workshop aimed to further develop the analysis framework and identify preliminary themes. Community researchers ($n = 6$) were divided into three groups, with the university researchers ($n = 2$) joining two of the groups. Transcripts were divided between these groups according to age; that is, one group received CSS and transcripts for the older generation (O = older), one for the middle generation (M = middle) and one for the younger generation (Y = younger). This was to highlight any generational differences in data. Each group received two interviews that they conducted, and four interviews conducted by other teams. Community researchers then used butcher's paper and post-it notes to cluster ideas into themes. The final step involved the CRs presenting their analysis to the whole group highlighting the core themes which were discussed and then formed the preliminary codebook.

The second stage of analysis involved an additional talanoa workshop with Elder CR and the university researchers. The Elder CR used the preliminary codebook and the remaining CSS with exemplar quotes to further extend and verify the themes. These themes were used to develop a free-form mind map that further facilitated analysis. This technique was used to facilitate the connection of concepts and to generate associations between ideas [51] and involved centralising the issue ('food') and then branching off related concepts and headings [52]. These branches were further subdivided and/or related back to other branches as analysis progressed, resulting in the development of the mind map. After this workshop, the two university researchers met to continue the mind map using colour coding to denote generational differences and asterisks to connect back to the themes identified in the first workshop.

The mind map and generated themes were presented (including in report form) to the SC and CR. The SC provided feedback on the wording, framing and ordering of the themes. Using multiple researchers and an analysis approach that involved multiple methods, including those that are highly visual, allowed for triangulation of data. A constant comparative approach that was managed by the university researchers ensured that data integrity was maintained.

2.5. Reflexivity

As is the practice in qualitative research and to acknowledge the shared expertise a reflexivity statement [53] is provided here. The authorship team comprises four members who identify as Pasifika (Akbar, Radclyffe, Santos, Mopio-Jane) of mixed cultural backgrounds. Four authors have been university trained in public health, nutrition, and anthropology. MM is an Elder of the Papua New Guinea community and a broadcaster for ethnic radio. Gallegos identifies as Australian and is a white, Anglo-Celtic, middle-class cis-female who has worked in academia for 15 years. A full listing of the SC and CR is provided in Supplementary Materials S4. At multiple points in this research project, community researchers were asked to reflect on their own positionality and on the process. This process was multi-layered with Elder and young researchers debriefing with each other, as well as CR reflecting with university researchers during all stages of the research.

2.6. Ethical Considerations

This research had ethical clearance from the Queensland University of Technology Human Research Ethics (#2000000381). The SC co-designed and reviewed all participant information and consent documents and all research processes to ensure their cultural integrity. All participants provided informed written or verbal consent. Any potential ethical issues arising during the interview were discussed allowing the participants to renegotiate informed consent based on the changing nature of the inquiry. Any cultural, spiritual and personal concerns about disclosure and anything in relation to cultural and religious beliefs were also discussed and clarified with participants before obtaining consent to disclose or use any sensitive information. For example, many Elders are secret keepers of specific cultural foods which have deep spiritual meaning. It was made clear that they were under no obligation to disclose this information.

3. Results

Fifty-nine community members were invited to participate, 15% ($n = 9$) declined and 17% ($n = 10$) were uncontactable or did not respond. Consequently, forty talanoa were undertaken with a total of 49 community members representing multiple Māori and Pasifika cultural identities (Table 1). Interviews were conducted with 34 individuals, 10 couples and there was one ($n = 5$) multi-Pasifika group talanoa. Approximately half (45% $n = 18$) were members of the older generation (>50 years), and the remaining participants were equally divided between the middle generation ($n = 11$, 30–50 years) and the younger generation ($n = 11$, <30 years). Two-thirds (67%) of participants identified as women.

From the talanoa, eight key themes emerged. In keeping with the Māori and Pasifika story-telling tradition, the themes are represented metaphorically and visually in Figure 3. Taro was chosen as it was described as a common staple food among most of those interviewed. Each theme is further discussed below with selected quotes. Quotes are designated by community group/cultural identification and generation (O; M; Y). Solutions and barriers to food insecurity will not be discussed in detail in this paper.

Table 1. Cultural identity as reported by participants in the interviews *.

Cultural Identity	Participants
Cook Islands	1
Federated States of Micronesia	1
Fiji	5
Kiribati	1
Kiribati/Australia	1
Māori/Pakeha	2
Māori	5
Māori/Cook Islands	1
Nauru/Australia	1
Niue	3
Papua New Guinea	4
Rotuma	1
Samoa	7
Samoa/Fiji	1
Samoa/Māori/Irish	1
Solomon Islands	2
Tokelau/Māori/Tuvalu/Cook Islander	2
Tokelau	2
Tonga	4
Tonga/Australia	2
Tuvalu	1
French Polynesian-Tahiti/Māori	1
Total	49

* We have reported the cultural identities as described by the participants. These include self-identification with a singular culture and with hyphenated identities dependent on the diasporic journey.

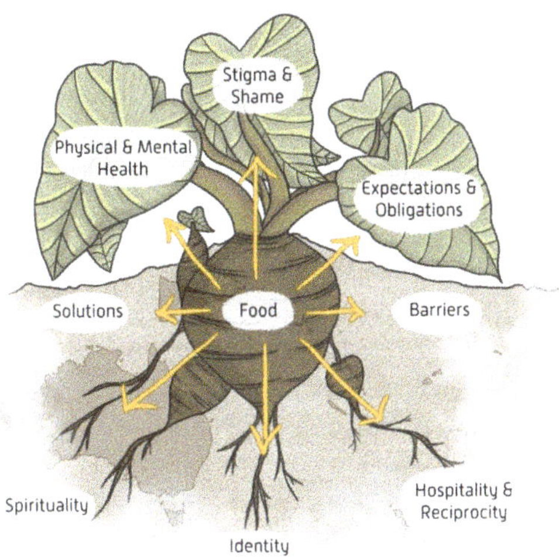

Figure 3. The taro plant illustrates eight key themes that emerged from the interviews which are shown to be all interconnected by food: identity, hospitality and reciprocity, and spirituality form the roots or foundation of the meaning of food for Māori and Pasifika peoples; solutions and barriers form the soil or structure from which these understandings and practices of food are nourished or restricted; physical and mental health, expectations and obligations, and stigma and shame form the leaves of the taro plant, representing the surface-level experiences and challenges of food insecurity. The diagram overlies a map of Australia, Aotearoa and some Pacific Islands to symbolise how our peoples' experiences of food insecurity may differ between diasporic and homeland communities but are simultaneously intertwined by socio-cultural and genealogical ties that transcend space and time.

It is important to note that these themes are not mutually exclusive, and they overlap and provide a rich description of food, identity and holistic health in Māori and Pasifika cultural groups living in south-east Queensland. The themes describe Māori and Pasifika as a collective but this in no way implies homogeneity across the island nations. Food as a cultural object is constantly evolving and its meaning varies not only from nation to nation but also from family to family and from individual to individual. We have where possible explored in depth the similarities and have attempted to highlight the differences. However, it has not been possible to give due acknowledgement to the deep and nuanced differences across cultural groups and generations.

3.1. Identity—Food Is Central to Cultural Identity and Connects People, Families and Communities

> *In Chuukese there is a saying 'food is bone' which explains the centrality of food to life, to cultural, familial and communal shared experiences. In the way that bone provides the skeletal structure of the human body, so food is the framework to a holistic life and shared social and cultural experiences" (O_Federated States of Micronesia).*

Food plays an important social and cultural role in all Pacific communities, often beyond nourishment. Forming a central component of Pasifika cultures and everyday living, food plays a vital role in keeping traditional practices alive. In traditional contexts, food is a means of maintaining societal norms and practices and affirms one's identity and sense of belonging. Food and its sharing is the social cement that provides a framework for shared social, cultural and communal experience and includes nurturing and sustaining individual, family and communal relationships, particularly intergenerationally for Māori and Pasifika peoples where food has and continues to be core to Pasifika concepts of love, reciprocity and respect.

> *It's sort of like a love language in ways . . . giving and receiving food. For example, when my father was in hospital and my mom was in town at the time, it was the fact that my friends showed up with food for us. That was a big deal for her (Y_Papua New Guinea).*

Every Pasifika culture has specific protocols on how certain types of foods are prepared, served and to who and by whom as a way of maintaining customs. It is through food that familial and community connections are forged, maintained and strengthened. Traditional foods are pivotal to this identity formation and are valued not only for their nutrition but also for learning about one's culture. Customary practices regarding how food is prepared, stored, distributed and consumed, is passed down through the generations. This includes the value of certain traditional foods (e.g., pig, taro/dalo) which signify respect and hierarchy. For example:

> *Certain foods are shared in social occasions or functions and have higher value such as ufi (dalo) and so these are shared or gifted with families and members as a sign of respect . . . For instance, different parts of fish or meat are given to certain important members of the family (usually the head or the eldest (O_Fiji).*

Food creates a 'safe cultural space' that provides opportunities for individuals to express themselves, reaffirming identity and a sense of belonging. Food unites cultural groups and communities. The sub-themes identified were food as: connection; a language of love; a way of honouring family and showing respect; a way to create culturally safe places; and as celebration. See Supplementary Materials S5 for exemplar quotes for sub-themes. The key role of traditional foods in keeping culture alive was highlighted:

> *We had one (underground oven) at the premises and one again at campsite. We have a number of youths [who] turned up and you can tell they thoroughly enjoyed the experience. This is how we cook back at home, how our ancestors cooked. It wasn't with an electric oven that we turn the thing on. It was with wood and rocks. (O_Niue).*

3.2. Hospitality and Reciprocity

Food plays a central role in demonstrating the key cultural tenets of hospitality, reciprocity, respect and humility as illustrated by the themes described below. Food is part of a communal system connecting the Māori and Pasifika diaspora: an offer to share food may simply be an expression of hospitality and reciprocity guided by spoken and unspoken rules and demonstrates cultural appreciation. As with identity, there is a collective understanding that the main purpose of contributing food for shared activities is to enable participation and inclusion of all members of a community. In this context, contributing food is taking responsibility and maintaining cultural traditions within a cultural system of support to preserve social order and hierarchy. Often this means not only bringing food but also eating the food that is prepared.

> *Hospitality is a big thing in our church, because it sort of creates that sense of connectedness again. So even at the moment through COVID, it's been a little bit tough, because we haven't been able to do like big connection groups like or big corporate gatherings. But when we do when we are able to have them, it's always kind of centred around food when we have that time to connect off afterwards (Y_ Māori).*

> *Hospitality, you know, people who would visit us, the first thing we would provide is a drink. And then after that, if they were staying longer then we would provide food and have a meal, for example. But food seems to be the thing that brings everyone together and then people share that hospitality, share that fellowship, [and] share the social side of things (O_ Rotuma).*

Reciprocity was conceptualized as 'nobody goes hungry' and 'there is always more than enough food'. Food is never offered sparingly as that would be deemed disrespectful and only the best is offered to guests and visitors. Whilst this can be detrimental to some families who are struggling, there is a communal culture of service with community members and families helping and supporting those in times of need.

> *Well, I think that, you know, what, some families, we're that that are not so well off, I think that is excellent, as well. terms of, I suppose that they have to make sacrifices and struggle a bit in some areas, because, you know, their country, they have to contribute to the community. But it's like a lot other Pacific Island communities here is very kind of communal type, you know, communal system, I suppose it's kind of like a bit of, you know, reciprocal behaviour that goes on as well. Because if you have a function as well, you know, for example, that the people from your church will help you out too . . . (O_ Niue).*

A formal practice of reciprocity is the tradition of 'Inati' (pronounced in-at-see), which is a shared concept practised for generations among Tokelauans, with similarities across other cultural groups. Inati implies that there will always be leftovers and plenty to go around. It means enough food has been provided and it is shared amongst everyone:

> *Inati basically means the sharing of food. It's contributing to each family . . . When they call the Inati, it is always boys. It's not voiced for the mothers or the fathers. It has always been said that, for us with the "moya tomaniti" (meaning our children), we say there is Inati happening at two o'clock this afternoon. Okay, so each family will go, one or two people will go to that Inati. . . . .So basically, what they do is they calculate how many people in that family and then they'll share accordingly. A family of three cannot have the same amount as a family of 10. So, they actually distribute accordingly (O_ Tokelau).*

While one side of reciprocity is in the giving, the other side is in accepting. To not accept what is offered is also considered disrespectful. However, some members of the younger generation believe this is changing.

> *So in my role, I've done home visits, things like that, if somebody offers me food I won't turn it down. It's rude. Where I'm from, it's rude. And I'm also seeing people who don't have lots of money who are in crisis. And if they're offering me something, I'll take it, because they're offering. That offer in itself means a lot when they haven't got enough (M_ Māori).*

I don't think it's not as, I guess, taboo to not eat someone's food that they've prepared for you. I guess if they prepared it specifically for you, then yes, you should have a plate or at least eat from the plate. There's no real expectation for you to finish it. But there is an expectation for you to keep serving (Y_ Kiribati).

3.3. Spirituality

Spirituality plays a pivotal role in Māori and Pasifika communities and is observed through religious or faith practices as well as metaphysical, mental and physical connections to land and ancestors. Food is sacred and therefore connected to spirituality. Food that is grown, collected, prepared, shared and consumed acknowledges an 'exchange of energy' or 'mana' between all elements of the physical, metaphysical and social spaces and these connections are essential to holistic health. In Australia, families who have migrated have adapted to their new environments, moving away from growing their own foods. As a result, this has contributed to cultural and spiritual disconnect from the process of growing and appreciating this 'exchange of energy through food and thus 'feeding our mana'.

We are very connected to the process of food, the way we work and toil the land, and so we appreciate food, you know. It's all connected and for us Pasifika you know food and the land, we [have to] thank God. It's a spiritual connection that we have to the land and how it gives back to us (M_Tonga).

I'd say mainly because it's one way that we show our appreciation or love. Because we don't live on monetary things, especially in the Islands. And a lot of it is because you put a lot of time and effort into growing it and cultivating it, and then into harvesting it. You've put all this time and energy into this and you're consuming it. So, you're consuming that energy for your body (Y_Kiribati).

From a religious perspective, food is honouring God with the first fruits of one's labour. Thus, sharing of food, for example, after a church service, affiliates with this purpose that 'mana' or strength comes from the ability to share and the ability to receive when one is vulnerable or hungry.

Not only the sacrament on Sunday, but the food that God provides for us every day is holy. And so therefore, we treat this as a provision that God has given to us to sustain our life, to develop us. Every holistic development of a human-being centres on this provision of God, and in your context this is food. So therefore, it is very important that food needs to be clean. Hygiene needs to come into all perspective of development, you know, gardening, growing, cooking, all those aspects; how to preserve them for the use of the family of the community, how to offer them in generosity to those who have come. And then they all are contained in God 'Mana'.... that's it (O_Rotuma).

3.4. Expectations and Obligations

Food for Māori and Pasifika communities is also centred around fulfilling complex obligations to the community and church. Expectations could include contributing by bringing food to an event or gathering, by contributing financially or through participation that is, food preparation, cleaning and/or serving. Being present and actively contributing in some way is valued and is seen as important in maintaining relationships and connections. However, if these obligations or duties are not fulfilled, there may be negative consequences for families or individuals within their communities. While participation was considered a form of contribution often the types, quality and quantity of food provided were taken as the definitive benchmark of the value of the contribution. In many communities the contribution was judged against the social norms that are spoken and unspoken:

I think there is definitely an expectation. I think there's an expectation that if you want to be part of a community and being an active member, that everyone must contribute. So if everyone's contributing food, then every member that turns up must contribute food. I think there's that cultural expectation there. So I think for those who choose not to be involved, it could be a fact that they just financially aren't able to contribute and then therefore cannot participate in those community events. And then that can kind of I guess impact our cultural identity, and not being able to be part of our cultural community can have such a massive impact (Y_Kiribati).

Mostly, because I give back. Yeah. So in my head, I'm like, it's not something that I would expect. It's something that I would do. More so out of obligation, rather than because I want to. Yeah. If that was the case, because it's kind of that notion of you scratch my back and I'll scratch yours. Which to me is like, I don't love it but, again because I think when I give or when I do it, there is the expectation and intention of not receiving. It's purely for the intention of giving (Y_Māori).

There is intergenerational transfer of cultural knowledge through food which is fundamental to maintaining identity, connectedness and traditional practices. For diasporic communities, there are, however, changing priorities for younger generations in relation to financial commitments, family structure (nuclear instead of extended families), with mixed marriages of partners not necessarily from the same cultural background, obligations and work and family commitments. For many of the younger generation, there is a shift away from traditional practices and how food is perceived in terms of access, preparation, buying, feeding and consumption.

I don't feel like it should always be an expectation. I think if I knew that my brother or sister wasn't working or were just making enough to provide for their own kids, that is expectation met. So it would be like if I'm hosting or having a big family dinner, the expectation is that they just turn up. I think it's just about finding that balance, because I think some of the expectations are old school. Really, really old school. And for me, I guess, the way that I live now isn't really applicable, or is not always applicable. So it's hard to sort of follow them (Y_Māori).

But I think that it's also dictated to by things such as changing priorities of young people, and the cost factor. For example, some young people, they just like a cup of tea, you know, some biscuits, sandwiches at their function … It is a lot easier to prepare. Usually when you're providing a Pacific Island type meal, it takes a long time to prepare. It can be very costly as well. Some of the younger ones, you know, got mortgages, they've got kids or got sports fees, and so forth (O_Niue).

3.4.1. Expectations from Family

For the Māori and Pasifika diaspora living in Australia and Aotearoa, it is common that expectations are placed on them to support extended families living in the same country or in their home Islands either financially or through other means. Pressures of not being able to fulfil such obligations can cause stress, conflict and discord within families and communities. It can also result in financial and material hardship for those living in Australia.

The view from the Islands with relatives is wealth but [they] do not realise that its more expensive to live here. I know there is an expectation for those of us overseas to contribute back home … One of the things that I've tried to do is explain to people back at home that some here and a significant number of Fijians and Pacific Islanders, they have to work two or three jobs to support their family. . . . . I know of people that are supporting their own families back home, they pay their electricity bill and everything else, including food." (O_Fiji, Samoa, Cook Islands, Māori).

3.4.2. Expectations from Churches

Churches in many Māori and Pasifika communities are an integral part of the spiritual and social nourishment of members. Churches play a pivotal role in Pasifika communities, providing pastoral care that includes spiritual, mental, physical and financial support. Being well connected to communities, church can also play a key role in providing cultural support by providing safe spaces for Indigenous languages, traditional dress and practices to be expressed. Through the offering of charitable services, churches provide a significant safety net for those in need. It is also a space where trust is established; thereby creating a system of support for those in need. Families facing hardship, who are connected to the church, will use this resource before seeking external help. Knowing who needs help through the 'coconut wireless' (a community communication network) ensures that the support can be reciprocated in time.

> *I think the role of the churches in its pastoral care is important that we look at people who do not have and to send food or think how do we help them? ... The money that God gives us as a gift can be channelled to other important things in the lives of the people because the church is a people. (O_Rotuma).*

However, different churches have different expectations of their members. For some participants, the expectations of church members to contribute equally, particularly to the financial viability of the organisation, potentially contributes to hardship and reduces the effectiveness of the safety net. This expectation was not necessarily experienced by the younger generation.

> *I found out why a lot of people weren't coming [to church]. Because they didn't have the koha (donation) (O_Niue).*

> *It can become a burden to the people that they are helping or an expectation from them to contribute ... It's a sad thing to say, but I think it's more than the forcefulness of it ... It's like I need you in the church to show that our church is growing, that's why they do what they do ... (O_Samoa).*

With churches being the predominant safety net, there can be challenges for families or individuals who are not part of a church or faith-based community. This is often the case for many newly arriving Māori and Pasifika individuals or families who are unfamiliar with the networks and may not know who to go to for support or help.

> *Unfortunately, there's also a lot of people that do not have that support. They don't attend any churches, they pretty much like live on their own really and are not involved in any of their respective communities and cultural communities. So they're the ones that will go without having family and social support around them (O_ Samoa).*

3.5. Stigma and Shame

There are two elements to this theme. The first is not being able to honour the expectations and obligations, outlined above, generating feelings of stigma, embarrassment and a sense of being judged. Families who are struggling may not be able to reciprocate or make financial or food contributions. Consequently, this could lead to not being able to connect to the wider community, resulting in further isolation and disengagement. Alternatively, the drive to honour these expectations can result in families going without food or using money set aside for rent or other core living expenses.

> *But I think in terms of anyone who shows up to an event where there's no food, it will be looked down upon. And probably classified as rude. It could just be they just don't have any money to be able to provide it and it's kind of a no-win situation sometimes which I find it really harsh in our culture (M_Samoa).*

> *It's sad, I see this a lot. When families haven't got anything to contribute, they stay home. Because sometimes it's a shame to go to our traditional thing when we know everyone else will be putting in something, and I don't have anything to put in. Because usually this time, they sometimes read out names. And if your name doesn't get read out, it can be a very shameful experience. So the family doesn't go. So what happens is more and more people don't go because [of] their financial (issues) (O_Niue).*

The second element is a culture of pride within many Māori and Pasifika communities to hide hardship. Shame is associated with not having food. Keeping family matters private or using family safety nets is a way to maintain social identity and family honour. This strong culture can create barriers and challenges to seeking help or support or disclosing to others about issues of food insecurity.

In my experience, the majority of the time this struggle was never expressed outside the family. So in the same way, as the individual might not talk about, might not bring their struggle to the community space . . . Maybe it's viewed as weakness, maybe as soon as they don't want to impose on the space, or the community (Y_Samoa).

I think there's still a lot . . . to learn, you know, when you're doing it tough to put your hand up. You know, again, it comes down to that sense of pride, where you don't want people to know or anyone else to know what you're going through. (O_Cook Islands).

Because I think it's a hard thing to go and ask for free food. Yeah, it's shame. Men are supposed to be supporting your family providing for your family and you and you can't even provide the basic of food on the table. I think it really took a kick in the gut kind of thing for them (Y_Tonga).

For many Pasifika men not being able to put food on the table represents a loss of their manhood and a sense of failure that they are unable to provide for their families which can impact family dynamics and relationships as well as food security. This may not be an issue for those living in the Pacific Islands where food has been in abundance and was able to be collected, freely, from the land and sea. This was, however, identified as an issue for those living in Australia.

Well, I can speak on behalf of men. And I know that for any normal man that can't put food on the table, it would be quite devastating. He would feel inadequate, and as if he wasn't doing his job properly. Especially if there's kids in the family. . . . . Because in Nauru, you don't really have a problem with providing food on the table, because you can just go and catch fish. And yeah, you can get everything you need off the land and off the sea (O_Nauru).

3.6. Physical and Mental Health

Health is an important issue for many Māori and Pasifika peoples living in Australia where there is a burden of chronic diseases and poor health status because of behaviours influenced by environmental (obesogenic environments—that is, easy access to energy-dense foods plus lack of time) and financial conditions. Participants identified feeding large families on small budgets, with adults working multiple jobs and with long hours, creating time poverty, as a particular issue. Providing cheap, energy-dense fast food was also considered an option for community events for those with fewer financial and time resources. Other key issues included lack of understanding about portion size, being mindful of eating healthily and the connections between food, nutrition and health.

So let's say you have two parents, and a large family. Between them working, then looking after the kids then coming home? The question is like, Where's the time to produce a good meal? Like where is the energy if they're running between looking after, maybe there's, a five-year-old and a sixteen-year-old, between them playing sports or juggling a lot of things? (Y_Samoa).

There may be the lack of knowledge and understanding the lack of skills to be able to know what to do with some of the different variety of foods. So one of the problems we have is too many portions, now we have too easy access. It's not really appreciated, we just buy a whole bunch, and we're eating copious amounts in every meal, because we went for the money and we're just trading with money. And so because of that disconnect of that cycle of development, growth, distribution, evening out that cycle in the Islands is almost a little bit different now (M_Tonga).

4. Discussion

Food security is an issue for the Māori and Pasifika diaspora living in south-east Queensland, but its ramifications go beyond the material provisioning of food and reach deeply into identity, belonging and community obligations. These in turn impact on health and wellbeing through disrupted relationships between food, environment, genealogical ties, cultural identity, resilience, reciprocity, respect and humanity [54,55]. The results from this research remind us that structural changes to improve access to and availability of healthy foods needs to take into consideration the expectations and obligations that bind communities together. The results highlight a range of points of discussion, but we want to focus on three key components; firstly, cultural food security is a salient consideration; secondly, that food security for Māori and Pasifika peoples is not an individual but a collective responsibility; and finally, food sovereignty approaches may provide more sustainable long-term solutions.

4.1. Cultural Food Security

Each domain of food security is influenced by culture from how food is grown, procured, prepared and shared [4]. The essentiality of cultural foods to food security acknowledges that cultural food security emphasizes the ability to reliably access important traditional foods, using traditional growing, harvesting and cooking techniques [43]. For many Māori and Pasifika peoples living in Australia, there is an accumulative impact of identifying as both Indigenous and migrant. Foodways and maintaining food security become imperative to building both a collective Pasifika identity as well as an individual Pacific Island nation identity. This may typically involve growing, preparing and sharing traditional foods in Australia, importing traditional foods from the Islands, maintaining to-and-fro movements between the diaspora and the Islands as well as the maintenance of these conduits by sending remittances and other support.

Food is one of the primary vehicles for social connectivity for the Māori and Pasifika diaspora. It is the materiality that underpins the spiritual, emotional, physical and social health of each member as well as the collective. Food is key to the building and maintenance of social capital [56]. Social capital (as social participation, networks and trust in others) is protective of migrant mental health, especially in those exposed to racism and discrimination [57,58] and can contribute to flourishing [59]. Flourishing is optimal functioning in emotional, psychological and social well-being—the opposite is languishing [59]. Cultural food security is therefore imperative for not only the maintenance of Maori and Pasifika community health, but is necessary to enable the communities to flourish.

Food insecurity in Māori and Pasifika communities is also contributing to the increased consumption of less healthy non-traditional options, foods that fill (are calorically dense) rather than nourish (are nutrient dense) [41,42]. There is a growing tension between the obligations and expectations associated with demonstrating hospitality and generosity with the need to limit portion sizes for health. The younger generations are providing leadership in finding compromises that fulfil community social obligations without compromising individual health.

4.2. Food Insecurity as a Collective Responsibility

The Food and Agriculture Organization (FAO) has recently extended the conceptualisation of food security to encompass two additional domains "agency" and "sustainability" [1]. The agency domain is relevant here in that it refers to the rights of individuals to determine their own food system. Agency as defined by Sen [60] is when a "person is free to do and achieve in pursuit of whatever goals and values her or she regards as important" (p. 203). While the domain stresses individuals and groups, the concept is still embedded in individualistic rather than collective conceptualisations.

Individualistic cultures which are pervasive in most countries with neoliberal governments focus primarily on the care of self and family with less reliance on others for support [61,62]. Personal goals and identity take precedence over the goals of society and social identity. Conversely, collectivist cultures, which encompass many Indigenous peoples, focus on communal responsibility, interdependence and collective survival [63]. Harmony and relationships and the health of the whole community rather than individuals are prioritised. There are indications that food security (FS) at the household level may be viewed differently from the perspective of collectivist versus individualistic cultures. Renzaho and Mellor [64] describe food sharing and providing for needy families not as a coping strategy but rather as a "cultural obligation and a prerequisite for cultural harmony and community cohesion" (p. 2).

A strength of the Māori and Pasifika diaspora is the reliance on the collective in a complex interplay between expectations and obligations to ensure that all members of the community have access to food. For many communities, this is mediated by faith-based organisations that play a central role in ensuring well-being. Religious identity has been shown to contribute to psychological wellbeing directly and indirectly through social connectedness and social support [65]. These networks are responsible for ensuring that pride is maintained, so families in need do not need to ask for help. The outward manifestations of food insecurity in collectivist cultures range from not being able to meet your obligations and maintain social connectivity through food, to going hungry. Food insecurity is therefore not an individual failing but potentially a failure of the collective network. The stigma and shame that accompanies food insecurity is consequently potentially experienced by not just the individual but reflects on the entire community. Stigma and shame associated with food insecurity are not unique to Māori and Pasifika communities. They are experienced almost universally by those who are struggling to put food on the table, and in particular by those who have resorted to using charitable food relief [66,67]. Dryland et al. [68] describe the maintenance of social identity as a key driver for women experiencing food insecurity. In these situations, women prioritised strategies that provided a public demonstration that their situations were not compromised. For Māori and Pasifika families, this can mean maintaining their obligations in terms of church gift-giving, hospitality, reciprocity, remittances to family in the Islands and contributions at social functions. While at home they may be struggling to put food on the table.

The depth of distress experienced by Māori and Pasifika communities could potentially impact on how household food insecurity is measured. The FAO has recently started to move away from using only undernourishment as an indicator for food insecurity towards an experience scale (Food Insecurity Experience Scale (FIES)) [69]. The FIES is an eight-question tool based on the measure used in the United States and asks at the individual or household level about experiences of food insecurity (from worry to hunger) due to lack of money or other resources. The same measure is being used across countries to allow cross-country comparisons. There is no data for the Oceania region apart from Australia and New Zealand [70]. Given the impact of climate change in the region and the ensuing transformation of food systems, understanding the level of food insecurity experienced at the community level is vital [71]. The FAO in a personal communication indicated "we face an issue with fielding the FIES in some Pacific Island states, from where we receive reports that the questions are perceived as offensive by members of local communities and we are looking for means to better adapt them to culture and context to preserve their efficacy in eliciting the signs of food insecurity" (FAO, personal communication 23 January 2019). Developing the appropriate measure for monitoring will require understanding the broader ramifications of food insecurity and working with communities to identify and explore the issue in depth.

4.3. Moving to a Food Sovereigty Approach

The more recent conceptualisation of food security to include agency and sustainability as key dimensions highlights the possibility of moving towards a food sovereignty approach. Food sovereignty implies a broader vision of food security, moving away from an individualised focus to one that disrupts and reimagines food systems so they are democratically managed and geographically specific [72,73]. Food sovereignty has been conceptualised within First Nations contexts, as having seven pillars: (1) focuses on food for people; (2) builds knowledge and skills; (3) works with nature; (4) values food providers; (5) localizes food systems; (6) puts control locally; and (7) allows that food is sacred [74].

This approach aligns with the conceptualization of food for Māori and Pasifika communities living in south-east Queensland. That is, food is deeply linked to identity, spirituality, culture and holistic health. Food sovereignty highlights that the human right to food encompasses access to food-producing resources such as land and water, which can be more challenging in urban environments and for those disconnected from traditional resources via migration [75]. Food sovereignty requires an understanding and disruption of the inherent power structures and a focus on self-determination and social justice to enable individuals and communities to define their own food system [76]. Food sovereignty is described as a process as much about creating connectivity as about creating autonomy [77]. A food sovereignty approach also links more tightly with the Sustainable Development Goals with a focus on ensuring the health of land and seas, the maintenance of livelihoods and the empowerment and education of people [78].

Currently, instead of collective food sovereignty approaches, individualised interventions such as emergency food provisioning or cooking and budgeting workshops are more pervasive. These provide a valuable food and skills safety net but do not address long-term structural barriers. Health programs, to date, have not focused on community responsibilities but have instead "promoted individualised empowerment discourses which dissuade a symbiotic understanding of self-in-relation to community and land, place-based agency and social–ecological resilience" [79] (p. 57). Māori and Pasifika communities already have powerful social networks, deep knowledge and skills about food systems and a spiritual imperative to build a collective food network. What will be key in developing any strategies for Māori and Pasifika communities is recognition that the community is not a homogenous community but rather a multiplicity of cultural backgrounds and identities. There are also generational changes –with younger generations potentially less strongly connected to existing community safety nets that older generations have relied on and drawn support from. Māori and Pasifika youth in the diaspora are struggling and reconnection to cultural identity, spirituality and community through food may offer one pathway to self-determination [33,80].

This work has identified that future research needs to focus on co-designing solutions using a food sovereignty approach that will enable communities to flourish. The impact of social issues such as structural violence and racism are under-explored determinants of food insecurity as are potential protective factors such as community cohesion and religiosity. Finally, for Māori and Pasifika peoples, ensuring that food insecurity can be measured and monitored in a respectful way is a priority to enable adequate resources to realign food systems for sustainable action.

4.4. Limitations

There are cultural differences between Māori and Pasifika communities and where possible these are highlighted but it has not been feasible in the context of this paper to explore the nuanced differences. The SC attempted to ensure representation from all Pacific Island nations, but we were unable to identify key informants from some of the smaller communities that may not be connected to the broader Pasifika diaspora. For example, there was no representation from the Marshall Islands or American Samoa. Some of the interviews took place at the height of the COVID-19 pandemic when many members of the community were in isolation. This situation had the potential of bringing food insecurity to the foreground due to the loss of jobs and income experienced by those who were ineligible for government payments. However, it also highlighted the resilience and strength of social networks that were mobilized to support community members. The data presented here is relevant within the context of Māori and Pasifika peoples living in south-east Queensland and may not be able to be generalized to the diaspora living in other locations. The conceptualization of food security as going beyond simple food provisioning for collectivist cultures could be relevant for other cultural groups.

5. Conclusions

Food has an important role in Māori and Pasifika communities in south-east Queensland and contributes significantly to collective identity, spirituality and holistic health. There are complex obligations and expectations involving food that are necessary for the maintenance of relationships and community. When foods (including cultural foods) are not available or accessible, individuals experience not only a loss of physical health but lose connections to their broader identity and their link to their ancestral homes. However, communities have demonstrated resilience through the maintenance of social, spiritual and cultural support networks. Churches and local community groups provide culturally safe and far-reaching safety nets for Māori and Pasifika families and individuals struggling to put food on the table.

Addressing food insecurity for collectivist cultures such as those of Māori and Pasifika peoples requires reframing its definition and scope to embrace food sovereignty approaches over individualised or nuclear family-focused methods. It requires a holistic appreciation of cultural food security whereby elements of identity, reciprocity, hospitality, mana and spirituality are all regarded as significant and interconnected. In this context, improved food security may be attained through co-creation and co-designing solutions which range from high-level actions that impact the social determinants of Māori and Pasifika peoples as well as practical solutions that utilise and strengthen existing social networks to produce and distribute affordable and nutritious food.

Supplementary Materials: The following supporting information can be downloaded at: https://www.mdpi.com/article/10.3390/nu14102020/s1. S1. SQRQ Checklist for Qualitative Research [81]. S2. Community Researchers (CR) interview guide. S3. Contact Summary Sheet (CSS) template used in data analysis [49]. S4. List of Steering Committee (SC) and Community Researchers (CR). S5. Exemplar quotes for identity sub-themes.

Author Contributions: Conceptualization, H.A. and D.G.; methodology, H.A., D.G. and D.S.; formal analysis, H.A., D.G., D.S., M.M.-J. and C.J.T.R.; writing—original draft preparation, H.A. and D.G.; writing—review and editing, D.G., H.A., M.M.-J., C.J.T.R. and D.S.; project administration, D.G. All authors have read and agreed to the published version of the manuscript.

Funding: This research received no external funding.

Institutional Review Board Statement: The study was conducted in accordance with the Declaration of Helsinki and approved by the Human Research Ethics Committee of the Queensland University of Technology (# 2000000381, approved 25 June 2020).

Informed Consent Statement: Informed consent was obtained from all subjects involved in the study.

Data Availability Statement: De-identified data is available on reasonable request to the researchers.

Acknowledgments: This project was supported by Woolworths Centre for Childhood Nutrition Research, Queensland University of Technology (QUT), and a broad network of Māori and Pasifika organisations, professionals, communities and families in southeast Queensland. The project was overseen by a dedicated SC formed in 2019 who were involved in every stage of the research. These members were Alvina Salemanesa, Aunty Violet Langan, Aunty Wynn Te Kani, Eddie Tavae, Daphne Santos, Iree Chow, Maureen Mopio-Jane, Nicole Alexander, Sarai Tafa and Radini Malo Vaimoso Semaia. Thanks also to Roman Kingi who joined the SC in 2021. The CR included seven community Elders, Heena Akbar, Nicole Alexander, Aunty Violet Langan, Maureen Mopio-Jane, Paul Ogil, Malo Semaia, Aunty Wynn Te Kani, and seven young community members, Jaelyn Biumaiwai, Iree Chow, Tahi Gray-Tuapawa, Tima Mata'afa, Charles Radclyffe, Daphne Santos and Sarai Tafa. Appreciation is extended to Nadia Baguley for graphic design.

Conflicts of Interest: D.G. was supported by the Queensland Children's Hospital Foundation through a philanthropic grant from Woolworths. H.A. and C.J.T.R. were supported by the Queensland University Technology through the Woolworths Centre for Childhood Nutrition Research. M.M.-J. is employed by Radio 4eb Ethnic Broadcasting Multicultural Association and is a member of the Papua New Guinea Federation Queensland Inc. D.S. is supported by the Good Start Program through Child and Youth Community Health Services, Children's Health Queensland. Funders or employees had no role in the design of the study; in the collection, analyses, or interpretation of data; in the writing of the manuscript, or in the decision to publish the results. The views expressed in this paper do not represent those of any individual organization.

References

1. HLPE. Food Security and Nutrition: Building a Global Narrative Towards 2030. In *A Report by the High Level Panel of Experts on Food Security and Nutrition of the Committee on World Food Security*; FAO: Rome, Italy, 2020.
2. Food and Agriculture Organization (FAO). *Coming to Terms with Terminology: Food Security, Nutrition Security, Food Security and Nutrition, Food and Nutrition Security*; Committee on World Food Security: Geneva, Switzerland, 2012.
3. Pollard, C.M.; Booth, S. Food insecurity and hunger in rich countries—It is time for action against inequality. *Int. J. Environ. Res. Public Health* **2019**, *16*, 1804. [CrossRef] [PubMed]
4. Alonso, B.E.; Cockx, L.; Swinnen, J. Culture and food security. *Glob. Food Sec.* **2018**, *17*, 113–127. [CrossRef]
5. Ramsey, R.; Giskes, K.; Turrell, G.; Gallegos, D. Food insecurity among Australian children: Potential determinants, health and developmental consequences. *J. Child Health Care* **2011**, *15*, 401–416. [CrossRef] [PubMed]
6. Thomas, M.M.C.; Miller, D.P.; Morrissey, T.W. Food insecurity and child health. *Pediatrics* **2019**, *144*, e20190397. [CrossRef]
7. Ramsey, R.; Giskes, K.; Turrell, G.; Gallegos, D. Food insecurity among adults residing in disadvantaged urban areas: Potential health and dietary consequences. *Public Health Nutr.* **2012**, *15*, 227–237. [CrossRef] [PubMed]
8. Dean, E.B.; French, M.T.; Mortensen, K. Food insecurity, health care utilization, and health care expenditures. *Health Serv. Res.* **2020**, *55* (Suppl. S2), 883–893. [CrossRef]
9. Banerjee, S.; Radak, T.; Khubchandani, J.; Dunn, P. Food insecurity and mortality in American adults: Results from the NHANES-linked mortality study. *Health Promot. Pract.* **2020**, *22*, 204–214. [CrossRef] [PubMed]
10. Gundersen, C.; Tarasuk, V.; Cheng, J.; De Oliveira, C.; Kurdyak, P. Food insecurity status and mortality among adults in Ontario, Canada. *PLoS ONE* **2018**, *13*, e0202642. [CrossRef]
11. Kirby, J.B.; Bernard, D.; Liang, L. The prevalence of food insecurity is highest among Americans for whom diet is most critical to health. *Diabetes Care* **2021**, *44*, e131–e132. [CrossRef]
12. Walker, R.J.; Garacci, E.; Ozieh, M.; Egede, L.E. Food insecurity and glycemic control in individuals with diagnosed and undiagnosed diabetes in the United States. *Prim. Care Diabetes* **2021**, *15*, 813–818. [CrossRef]
13. Jones, A.D. Food insecurity and mental health status: A global analysis of 149 countries. *Am. J. Prev. Med.* **2017**, *53*, 264–273. [CrossRef] [PubMed]
14. Elgar, F.J.; Pickett, W.; Pförtner, T.-K.; Gariépy, G.; Gordon, D.; Georgiades, K.; Davison, C.; Hammami, N.; MacNeil, A.H.; Azevedo Da Silva, M.; et al. Relative food insecurity, mental health and wellbeing in 160 countries. *Soc. Sci. Med.* **2021**, *268*, 113556. [CrossRef] [PubMed]
15. Shafiee, M.; Vatanparast, H.; Janzen, B.; Serahati, S.; Keshavarz, P.; Jandaghi, P.; Pahwa, P. Household food insecurity is associated with depressive symptoms in the Canadian adult population. *J. Affect. Disord.* **2021**, *279*, 563–571. [CrossRef] [PubMed]
16. Men, F.; Elgar, F.J.; Tarasuk, V. Food insecurity is associated with mental health problems among Canadian youth. *J. Epidemiol. Community Health* **2021**, *75*, 741. [CrossRef] [PubMed]
17. Gallegos, D.; Eivers, A.; Sondergeld, P.; Pattinson, C. Food insecurity and child development: A state-of-the-art review. *Int. J. Environ. Res. Public Health* **2021**, *18*, 8990. [CrossRef]
18. Nolan, M.; Rikard-Bell, G.; Mohsin, M.; Williams, M. Food insecurity in three socially disadvantaged localities in Sydney, Australia. *Health Promot. J. Aust.* **2006**, *17*, 247–253. [CrossRef]
19. Seivwright, A.N.; Callis, Z.; Flatau, P. Food insecurity and socioeconomic disadvantage in Australia. *Int. J. Environ. Res. Public Health* **2020**, *17*, 559. [CrossRef]
20. Kerz, A.; Bell, K.; White, M.; Thompson, A.; Suter, M.; McKechnie, R.; Gallegos, D. Development and preliminary validation of a brief household food insecurity screening tool for paediatric health services in Australia. *Health Soc. Care Community* **2021**, *29*, 1538–1549. [CrossRef]
21. Butcher, L.M.; O'Sullivan, T.A.; Ryan, M.M.; Lo, J.; Devine, A. Utilising a multi-item questionnaire to assess household food security in Australia. *Health Promot. J. Aust.* **2019**, *30*, 9–17. [CrossRef]
22. Mansour, R.; John, J.R.; Liamputtong, P.; Arora, A. Prevalence and risk factors of food insecurity among Libyan migrant families in Australia. *BMC Public Health* **2021**, *21*, 2156. [CrossRef]
23. McKenzie, H.J.; McKay, F.H. Food as a discretionary item: The impact of welfare payment changes on low-income single mother's food choices and strategies. *J. Poverty Soc. Justice* **2017**, *25*, 35–48. [CrossRef]
24. Temple, J.B.; Booth, S.; Pollard, C.M. Social assistance payments and food insecurity in Australia: Evidence from the Household Expenditure Survey. *Int. J. Environ. Res. Public Health* **2019**, *16*, 455. [CrossRef] [PubMed]
25. Smith, M.D.; Rabbitt, M.P.; Coleman-Jensen, A. Who are the world's food insecure? New evidence from the Food and Agriculture Organization's Food Insecurity Experience Scale. *World Dev.* **2017**, *93*, 402–412.
26. Chilton, M.; Knowles, M.; Bloom, S.L. The intergenerational circumstances of household food insecurity and adversity. *J. Hunger Environ. Nutr.* **2017**, *12*, 269–297. [CrossRef] [PubMed]
27. Chilton, M.; Knowles, M.; Rabinowich, J.; Arnold, K.T. The relationship between childhood adversity and food insecurity: 'It's like a bird nesting in your head'. *Public Health Nutr.* **2015**, *18*, 2643–2653. [CrossRef] [PubMed]

28. Odoms-Young, A.; Bruce, M.A. Examining the impact of structural racism on food insecurity: Implications for addressing racial/ethnic disparities. *Fam. Community Health* **2018**, *41* (Suppl. S2), S3–S6. [CrossRef]
29. Phojanakong, P.; Brown Weida, E.; Grimaldi, G.; Lê-Scherban, F.; Chilton, M. Experiences of racial and ethnic discrimination are associated with food insecurity and poor health. *Int. J. Environ. Res. Public Health* **2019**, *16*, 4369. [CrossRef]
30. Conroy, A.A.; Cohen, M.H.; Frongillo, E.A.; Tsai, A.C.; Wilson, T.E.; Wentz, E.L.; Adimora, A.A.; Merenstein, D.; Ofotokun, I.; Metsch, L. Food insecurity and violence in a prospective cohort of women at risk for or living with HIV in the US. *PLoS ONE* **2019**, *14*, e0213365. [CrossRef]
31. Batley, J. *What Does the 2016 Census Reveal about Pacific Islands Communities in Australia?* Australian National University: Canberra, Australia, 2017.
32. McGavin, K.J.T.C.P. Being "nesian": Pacific Islander identity in Australia. *Contemp. Pac.* **2014**, *26*, 126–154. [CrossRef]
33. Durham, J.; Fa'avale, N.; Fa'avale, A.; Ziesman, C.; Malama, E.; Tafa, S.; Taito, T.; Etuale, J.; Yaranamua, M.; Utai, U.; et al. The impact and importance of place on health for young people of Pasifika descent in Queensland, Australia: A qualitative study towards developing meaningful health equity indicators. *Int. J. Equity Health* **2019**, *18*, 81. [CrossRef]
34. Enari, D.; Faleolo, R. Pasifika collective well-being during the COVID-19 crisis: Samoans and Tongans in Brisbane. *J. Indig. Soc. Dev.* **2020**, *9*, 110–126.
35. Faleolo, R. Pasifika diaspora connectivity and continuity with Pacific homelands: Material culture and spatial behaviour in Brisbane. *Aust. J. Anthropol.* **2020**, *31*, 66–84. [CrossRef]
36. Shepherd, S.M.; Ilalio, T. Maori and Pacific Islander overrepresentation in the Australian criminal justice system—What are the determinants? *J. Offender Rehabil.* **2016**, *55*, 113–128. [CrossRef]
37. Rodriguez, L. The subjective experience of Polynesians in the Australian health system. *Health Soc. Rev.* **2013**, *22*, 411–421. [CrossRef]
38. Rodriguez, L.; George, J.R.; McDonald, B. An inconvenient truth: Why evidence-based policies on obesity are failing Māori, Pasifika and the Anglo working class. *Kōtuitui N. Z. J. Soc. Sci. Online* **2017**, *12*, 192–204. [CrossRef]
39. Hill, K.; Ward, P.; Grace, B.S.; Gleadle, J. Social disparities in the prevalence of diabetes in Australia and in the development of end stage renal disease due to diabetes for Aboriginal and Torres Strait Islanders in Australia and Maori and Pacific Islanders in New Zealand. *BMC Public Health* **2017**, *17*, 802. [CrossRef]
40. Queensland Government; Department of Health. *Queensland Health's Response to Pacific Islander and Māori Health Needs Assessment*; Divison of the Chief Health Officer, Queensland Health: Brisbane, Australia, 2011.
41. Perkins, K.C.; Ware, R.; Felise Tautalasoo, L.; Stanley, R.; Scanlan-Savelio, L.; Schubert, L. Dietary habits of Samoan adults in an urban Australian setting: A cross-sectional study. *Public Health Nutr.* **2016**, *19*, 788–795. [CrossRef]
42. Gallegos, D.; Do, H.; Gia To, Q.; Vo, B.; Goris, J.; Alraman, H. Eating and physical activity behaviours among ethnic groups in Queensland, Australia. *Public Health Nutr.* **2020**, *23*, 1991–1999. [CrossRef]
43. Power, E.M. Conceptualizing food security for Aboriginal people in Canada. *Can. J. Public Health* **2008**, *99*, 95–97. [CrossRef]
44. Hamer, P. "Unsophisticated and unsuited": Australian barriers to Pacific Islander immigration from New Zealand. *Political Sci.* **2014**, *66*, 93–118. [CrossRef]
45. Smith, A. The fiscal impact of the trans-Tasman travel arrangement: Have the costs become too high? *Aust. Tax Forum* **2018**, *33*, 249–279.
46. Ash, J.; Campbell, J. Climate change and migration: The case of the Pacific Islands and Australia. *J. Pac. Stud.* **2016**, *36*, 53–72.
47. Chilton, M.; Rose, D. A rights-based approach to food insecurity in the United States. *Am. J. Public Health* **2009**, *99*, 1203–1211. [CrossRef] [PubMed]
48. Vaioleti, T.M. Talanoa research methodology: A developing position on Pacific research. *Waikato J. Educ.* **2006**, *12*, 21–34. [CrossRef]
49. Miles, M.B.; Huberman, A.M.; Saldaña, J. *Qualitative Data Analysis: A Methods Sourcebook*; SAGE: Los Angeles, CA, USA, 2020.
50. Braun, V.; Clarke, V. *Thematic Analysis: A Practical Guide*; SAGE: Newcastle upon Tyne, UK, 2021.
51. Davies, M. Concept mapping, mind mapping and argument mapping: What are the differences and do they matter? *Higher Educ.* **2011**, *62*, 279–301. [CrossRef]
52. Vindrola-Padros, C.; Johnson, G.A. Rapid techniques in qualitative research: A critical review of the literature. *Qual. Health Res.* **2020**, *30*, 1596–1604. [CrossRef]
53. Gallegos, D., Chilton, M.M. Re-evaluating expertise: Principles for food and nutrition security research, advocacy and solutions in high-income countries. *Int. J. Environ. Res. Public Health* **2019**, *16*, 561. [CrossRef]
54. McKerchar, C.; Bowers, S.; Heta, C.; Signal, L.; Matoe, L. Enhancing Māori food security using traditional kai. *Global Health Promot.* **2015**, *22*, 15–24. [CrossRef]
55. Moeke-Pickering, T.; Heitia, M.; Heitia, S.; Karapu, R.; Cote-Meek, S. Understanding Maori food security and food sovereignty issues in Whakatane. *Mai J.* **2015**, *4*, 29–42.
56. Porter, R.; McIlvaine-Newsad, H. Gardening in green space for environmental justice: Food security, leisure and social capital. *Leisure/Loisir* **2013**, *37*, 379–395. [CrossRef]
57. Lecerof, S.S.; Stafström, M.; Westerling, R.; Östergren, P.-O. Does social capital protect mental health among migrants in Sweden? *Health Promot. Int.* **2015**, *31*, 644–652. [CrossRef] [PubMed]
58. Du Plooy, D.R.; Lyons, A.; Kashima, E.S. Social capital and the well-being of migrants to Australia: Exploring the role of generalised trust and social network resources. *Int. J. Intercult. Relat.* **2020**, *79*, 1–12. [CrossRef]

59. Keyes, C.L.M.; Annas, J. Feeling good and functioning well: Distinctive concepts in ancient philosophy and contemporary science. *J. Posit. Psychol.* **2009**, *4*, 197–201. [CrossRef]
60. Sen, A. Well-being, agency and freedom: The Dewey lectures 1984. *J. Philos.* **1985**, *82*, 169–221. [CrossRef]
61. Parker, A.G.; Grinter, R.E. Collectivistic health promotion tools: Accounting for the relationship between culture, food and nutrition. *Int. J. Hum.-Comput. Stud.* **2014**, *72*, 185–206. [CrossRef]
62. Lanham, A.; Lubari, E.; Gallegos, D.; Radcliffe, B. Health promotion in emerging collectivist communities: A study of dietary acculturation in the South Sudanese community in Logan City, Australia. *Health Promot. J. Aust.* **2022**, *33*, 224–231. [CrossRef]
63. Hofstede, G.; Hofstede, G.J.; Minkov, M. *Cultures and Organizations*; McGraw-Hill Professional Publishing: New York, NY, USA, 2010.
64. Renzaho, A.M.N.; Mellor, D. Food security measurement in cultural pluralism: Missing the point or conceptual misunderstanding? *Nutrition* **2010**, *26*, 1–9. [CrossRef]
65. Hashemi, N.; Marzban, M.; Sebar, B.; Harris, N. Religious identity and psychological well-being among middle-eastern migrants in Australia: The mediating role of perceived social support, social connectedness, and perceived discrimination. *Psychol. Relig. Spiritual.* **2020**, *12*, 475. [CrossRef]
66. Booth, S.; Begley, A.; Mackintosh, B.; Kerr, D.A.; Jancey, J.; Caraher, M.; Whelan, J.; Pollard, C.M. Gratitude, resignation and the desire for dignity: Lived experience of food charity recipients and their recommendations for improvement, Perth, Western Australia. *Public Health Nutr.* **2018**, *21*, 2831–2841. [CrossRef]
67. Osei-Kwasi, H.A.; Nicolaou, M.; Powell, K.; Holdsworth, M. "I cannot sit here and eat alone when I know a fellow Ghanaian is suffering": Perceptions of food insecurity among Ghanaian migrants. *Appetite* **2019**, *140*, 190–196. [CrossRef]
68. Dryland, R.; Carroll, J.-A.; Gallegos, D. Moving beyond Coping to Resilient Pragmatism in Food Insecure Households. *J. Poverty* **2021**, *25*, 269–286. [CrossRef]
69. Food and Agriculture Organization. Voices of the Hungry: The Food Insecurity Experience Scale. Available online: https://www.fao.org/in-action/voices-of-the-hungry/fies/en/ (accessed on 8 March 2022).
70. FAO; UNICEF; WFP; WHO. The State of Food Security and Nutrition in the World 2021. In *Transforming Food Systems for Food Security, Improved Nutrition and Affordable Healthy Diets for All*; FAO: Rome, Italy, 2020.
71. Savage, A.; Bambrick, H.; Gallegos, D. Climate extremes constrain agency and long-term health: A qualitative case study in a Pacific Small Island Developing State. *Weather Clim. Extrem.* **2021**, *31*, 100293. [CrossRef]
72. Dennis, M.K.; Robin, T. Healthy on our own terms. *Crit. Diet.* **2020**, *5*, 4–11. [CrossRef]
73. Richmond, C.; Kerr, R.B.; Neufeld, H.; Steckley, M.; Wilson, K.; Dokis, B. Supporting food security for Indigenous families through the restoration of Indigenous foodways. *Can. Geogr.* **2021**, *65*, 97–109. [CrossRef]
74. Food Secure Canada. What is Food Sovereignty? Available online: https://foodsecurecanada.org/who-we-are/what-food-sovereignty (accessed on 8 March 2022).
75. Jarosz, L. Comparing food security and food sovereignty discourses. *Dialogues Hum. Geogr.* **2014**, *4*, 168–181. [CrossRef]
76. Weiler, A.M.; Hergesheimer, C.; Brisbois, B.; Wittman, H.; Yassi, A.; Spiegel, J.M. Food sovereignty, food security and health equity: A meta-narrative mapping exercise. *Health Policy Plan.* **2014**, *30*, 1078–1092. [CrossRef]
77. Shattuck, A.; Schiavoni, C.M.; VanGelder, Z. Translating the politics of food sovereignty: Digging into contradictions, uncovering new dimensions. *Globalizations* **2015**, *12*, 421–433. [CrossRef]
78. United Nations. The Sustainable Development Goals. Available online: https://sdgs.un.org/goals (accessed on 8 March 2022).
79. Ray, L.; Burnett, K.; Cameron, A.; Joseph, S.; LeBlanc, J.; Parker, B.; Recollet, A.; Sergerie, C. Examining Indigenous food sovereignty as a conceptual framework for health in two urban communities in Northern Ontario, Canada. *Glob. Health Promot.* **2019**, *26* (Suppl. S3), 54–63. [CrossRef]
80. Durham, J.; Tafa, S.; Etuale, J.; Nosa, V.; Fa'avale, A.; Malama, E.; Yaranamua, M.; Taito, T.; Ziesman, C.; Fa'avale, N. Belonging in multiple places: Pasifika young peoples' experiences of living in Logan. *J. Youth Stud.* **2022**, 1–17. [CrossRef]
81. O'Brien, B.C.; Harris, I.B.; Beckman, T.J.; Reed, D.A.; Cook, D.A. Standards for reporting qualitative research: A synthesis of recommendations. *Acad. Med.* **2014**, *89*, 1245–1251. [CrossRef]

MDPI
St. Alban-Anlage 66
4052 Basel
Switzerland
Tel. +41 61 683 77 34
Fax +41 61 302 89 18
www.mdpi.com

Nutrients Editorial Office
E-mail: nutrients@mdpi.com
www.mdpi.com/journal/nutrients